Mini Bajaj

Treatment of phenolic wastewater in suspended and fixed bed bioreactors

Karlsruher Berichte zur Ingenieurbiologie
Band 45

Institut für Ingenieurbiologie und Biotechnologie des
Abwassers, Universität Karlsruhe (TH)

Herausgeber: Prof. Dr. rer. nat. J. Winter

Treatment of phenolic wastewater in suspended and fixed bed bioreactors

by
Mini Bajaj

universitätsverlag karlsruhe

Dissertation, genehmigt von der Fakultät für Bauingenieur-, Geo- und
Umweltwissenschaften der Universität Fridericiana zu Karlsruhe (TH), 2008
Referenten: Prof. Dr. rer. nat. J. Winter, Prof. Dr.-Ing. E.h. H. H. Hahn, PhD

Impressum

Universitätsverlag Karlsruhe
c/o Universitätsbibliothek
Straße am Forum 2
D-76131 Karlsruhe
www.uvka.de

Universitätsverlag Karlsruhe 2009
Print on Demand

Gedruckt mit Unterstützung
des Deutschen Akademischen Austauschdienstes

ISSN: 0172-8709
ISBN: 978-3-86644-316-7

Treatment of phenolic wastewater in suspended and fixed bed bioreactors

Zur Erlangung des akademischen Grades eines

DOKTOR-INGENIEURS

von der Fakultät für

Bauingenieur-, Geo- und Umweltwissenschaften

der Universität Fridericiana zu Karlsruhe (TH)

genehmigte

DISSERTATION

Von

Mini Bajaj, M.Sc., M.Tech

aus Amritsar, Indien

<table>
<tr><td>Tag der mündlichen Prüfung</td><td>05.11.08</td></tr>
<tr><td>Hauptreferent</td><td>Prof. Dr. rer. nat. J. Winter</td></tr>
<tr><td>Korreferent</td><td>Prof. Dr. Ing. E.h. H.H. Hahn, PhD</td></tr>
</table>

Karlsruhe 2008

About Thesis

This dissertation is based upon five journal manuscripts discussing the biodegradation of phenol and 2-chlorophenol and one article published in a conference proceeding. These are as following:

I   Bajaj M., Gallert C. and Winter J. (2008). Biodegradation of high phenol containing synthetic wastewater by an aerobic fixed bed reactor. Bioresource Technol. 99 (17) : 8376-8381.

II   Bajaj M., Gallert C. and Winter J. (2008). Effect of co-substrates on aerobic phenol biodegradation by acclimatized and non-acclimatized enrichment cultures. Eng. Life Sci. 8 (2) : 125-131.

III   Bajaj M., Gallert C. and Winter J. (2008). Anaerobic biodegradation of high strength 2-chlorophenol-containing synthetic wastewater in a fixed bed reactor. Chemosphere. 73 (5) : 705-710

IV   Bajaj M., Gallert C. and Winter J. (2008). Treatment of phenolic wastewater in an anaerobic fixed bed reactor (AFBR) - Recovery after shock loading. J. Hazad. Mater. doi:10.1016/j.jhazmat.2008.06.027.

V   Bajaj M., Gallert C. and Winter J. (2008). Biokinetics of phenol degrading culture under aerobic batch conditions. (Communicated).

VI   Bajaj M., Winter J. and Gallert C. (2006). Response of shock loading on phenolic wastewater biodegradation in an anoxic suspended bed reactor. In proceedings: 2nd International conference on environmental research and assessment, Bucharest 5-8th, Oct.

All important results from the above papers are included in this document. Beside these papers, some more results upon phenol biodegradation in suspended bed reactors have been added to the thesis. The aim of this thesis is to provide the reader with a self-contained account of research. It is documented in a way to maximize accessibility to the reader while still providing enough explanation of the work done. In Chapter 5 summary of this document is also provided in German to illustrate the main results.

Acknowledgments

I express my heartiest gratitude to my supervisor Prof. Dr. habil. rer. nat. Josef Winter for the countless discussions, useful suggestions, encouragement through out my research period and critical review of the manuscript. Beside theoretical directions, I am very grateful for his support in the laboratory too. I thank em. Prof. Dr.-Ing. E.h. Harmann H. Hahn Ph.D. for being my co-referee and writing recommendation letters to DAAD for my scholarship extensions. Extended and special thanks go to PD Dr. rer. nat. Claudia Gallert for her interest in discussions, suggestions, constructive criticisms and motivations which helped me to improve this work. I am thankful to Prof. Dr.-Ing. Dr. h.c. mult. Franz Nestmann, Dean of the Faculty of Civil Engineering-, Geo- and Environmental Sciences for accepting my candidature. PD Dr.-Ing. Thomas Neumann is also gratefully acknowledged for his contribution as a member of my examination committee.

Many thanks also go to all my former and present colleagues at the Institute of Ingineurbiologie und Biologie des Abwassers for their support in the laboratory- especially by Frau Renate Anschütz, for the discussions during institute seminars as well as for the chit chats over "Tee, Kaffee und Kuchen". I thank extra Dr. rer. nat. Stephan Bathe for his useful hints. Martina Adamek and Dominik Freikowski are also acknowledged for German translations of summary and abstract of this thesis. I am thankful to Frau Rita Seith for always helping me in official matters.

Best thanks to DAAD- The German Academic Exchange Service for providing me the financial assistance for doing research and giving me a chance to learn about Germany, its language and people. I would also like to mention Ministry of Human Resources and Development, Govt. of India, who together with DAAD selected me for the doctoral fellowship.

Last but not least, I would like to extend my gratitude to my parents especially my mother for her unconditional love and constant prayers for my well being and successful finish of this work. They were worried for my research more than me. Heartiest thanks to my sister for her belief in me and for her never ending support which have often helped me a great deal through the ups and downs of research. Deserved thanks are for my brother for letting me to forget about all worries and amusing me with his talks.

Abstract

Phenol and chlorophenols are among the most important class of raw materials in chemical industry. These compounds also list among priority pollutants as they are highly toxic and bactericidal substances. Therefore treatment of industrial wastewaters containing concentrated phenolic compounds is obligatory. The main problem in treating phenol or chlorophenol containing wastewater is the toxicity it exerts to the microbial flora in biological treatment plants. Because of this toxicity there is a danger of partial or complete treatment plant failure, when the microbial flora is not adapted to phenol concentrations in the influent. The purpose of the present study is to adapt the microbial flora of domestic sewage sludge to phenol and 2-chlorophenol (2-CP) at high concentration under continuous feeding conditions for long time periods and to study the response of suspended and fixed bed systems to transient loading and operation. The studies of phenol degradation were carried out in aerobic, anaerobic and anoxic suspended bed reactors with the effluents from aerobic and anaerobic reactors, respectively, which were used as seed inoculum for the start up of fixed bed reactors. Anaerobic and aerobic sludges grown on glucose for long time periods in the suspended bed reactors were sensitive to the increase in phenol concentration in the synthetic wastewater. The performance of anoxic suspended bed reactors in terms of high organic loading was much better than that of aerobic or anaerobic suspended bed reactors. This reactor had an OLR of average 6 g COD/l·d and could sustain organic loads up to 10 g COD/l·d; however the presence of high concentrations of phenol in the influent was not favourable. The fixed bed reactors were able to sustain high pollutant concentrations in synthetic wastewater i.e. up to 3.7 to 4.7 g/l of phenol under anaerobic and aerobic conditions and 2.6 g/l of 2-CP in an anaerobic reactor. A high phenol removal rate of 2.93 g/l·d was achievable in the aerobic fixed bed reactor at a phenol/COD ratio of 0.8. Thus, it was feasible to use aerobic biological treatment of wastewater with phenol concentrations as high as 5 g/l. The negative effect of absence of nitrogenous substances in the wastewater on the phenol acclimatised aerobic flora was reversible, when the previous conditions were maintained. The aerobic batch assays of phenol degradation with co-substrate addition suggested different degradation behaviour for different inocula. The data of batch assays carried out with the sludge of the aerobic fixed bed reactor fitted very well with the Haldane kinetic model, giving a non linear regression with a correlation coefficient of 0.987 and an inhibition constant K_i for phenol of 648.13 mg/l, higher than in most previous studies with mixed cultures. In the anaerobic fixed bed reactor, a steady state phenol removal of 1.48 g phenol/l·d at an influent phenol concentration of 3.7 g/l, was observed. A phenol shock load by 0.94 g/l phenol increment in concentration was fatal for the stability of reactor. Partial recovery of the reactor was possible only when no feed was provided to the reactor for one month and the phenol concentration was kept at 0.19 g/l at restart.

For dechlorination studies of monochlorinated phenol a mixed consortium from an anaerobic reactor treating municipal sewage sludge was acclimatised to 2-CP under batch conditions and then the supernatant was used to charge the continuous fixed bed reactor. A high 2-CP degradation rate of 0.87 g/l·d in an anaerobic fixed bed reactor was obtainable. The inhibitory effects due to high chlorophenol loading were recoverable if such loading was only of short term and previous loading conditions were following. The ratio of chloride released per unit of 2-CP removed indicated the successful dechlorination of 2-CP. Peptone and yeast extract in the feed could be replaced by NH_4NO_3 and a defined vitamin solution, without a permanent negative effect on the microbial flora or reactor operation. Thus the continuous biodegradation of deleterious aromatic compounds in concentrated wastewater is possible if the microbial flora is adapted to the pollutant in small concentration increments in a fixed bed system, which is ideal to support an adapted microbial flora for long time periods and minimise the unfavourable microbial washouts.

Kurzfassung

Phenol und Chlorphenol gehören zur wichtigsten Klasse von Rohmaterialien in der chemischen Industrie. Diese Stoffe sind aufgrund ihrer hohen Toxizität und bakteriziden Wirkung als wichtige Schadstoffe aufgeführt. Deshalb ist die Behandlung industriellen Abwassers, welches hohe Konzentrationen an phenolischen Verbindungen enthält, notwendig. Das größte Problem in der Behandlung phenol- oder chlorphenolhaltigem Abwasser ist die toxische Wirkung hoher Konzentrationen auf die mikrobielle Flora der biologischen Behandlungsanlagen. Aufgrund dieser Toxizität besteht die Gefahr des teilweisen oder vollständigen Ausfalls einer Abwasserbehandlungsanlage, wenn die mikrobielle Flora nicht an die Phenolkonzentrationen im Zulauf angepasst ist. Das Ziel dieser Studie ist die mikrobielle Flora kommunalen Abwassers oder von Klärschlamm an hohe Konzentrationen von Phenol und 2-Chlorphenol (2-CP) unter kontinuierlichen Arbeitsbedingungen über lange Zeiträume anzupassen und Leistungsfähigkeit von Schwebebett- und Festbettsystemen auf schwankende Beladung und im regelmäßigen Betrieb zu erforschen. Die Studien über den Phenolabbau wurden in aeroben, anaeroben und anoxischen Schwebebettreaktoren durchgeführt. Die Abläufe der aeroben und anaeroben Reaktoren wurden als Inokulum für die Inbetriebnahme der Festbettreaktoren verwendet. Anaerobe und aerobe Schlämme, die über lange Zeiträume in den Reaktoren mit Glukose angezogen wurden, reagierten empfindlich auf die steigende Phenolkonzentration in synthetischem Abwasser. Die Leistung der anoxischen Schwebebettreaktoren in Bezug auf eine hohe organische Beladung war viel besser als bei den aeroben oder anaeroben Schwebebettreaktoren. Diesem Reaktor wurde eine durchschnittliche OLR von 6 g COD/l·d zugeführt. Er konnte eine organische Beladung von bis zu 10g COD/l·d vertragen, jedoch wirkte sich eine hohe Konzentration an Phenol im Zufluss ungünstig auf den CSB-Abbau aus. Die Festbettreaktoren konnten hohe Schadstoffkonzentrationen in synthetischem Abwasser aushalten, z.B. bis zu 3,7 und 4,7 g/l Phenol unter anaeroben und aeroben Bedingungen und 2,6 g/l 2-CP in einem anaeroben Reaktor. Eine hohe Phenolabbaurate von 2,93 g/l·d konnte in dem aeroben Festbettreaktor mit einem Phenol/COD-Verhältnis von 0,8 erzielt werden. Somit war eine biologische Abwasserbehandlung mit Phenolkonzentrationen bis zu 5 g/l möglich. Der Effekt der fehlenden stickstoffhaltigen Substanzen im Abwasser auf die aerobe, an Phenol angepasste Flora war reversibel, wenn die vorausgegangenen

Bedingungen aufrecht gehalten wurden. Die aeroben batch Ansätze zum Phenolabbau mit Zugabe von Co-Substraten zeigten verschiedene Verhaltensweisen für unterschiedliche Inokuli an. Die Daten der batch Ansätze, welche mit Schlamm der aeroben Festbettreaktoren ausgeführt wurden, passten sehr gut zu dem kinetischen Modell von Haldane, da sie eine nicht-lineare Regression mit einem korrelierendem Koeffizient von 0,987 und einer Hemmkonstante K_i für Phenol von 648,13 mg/l ergab, was höher ist als in den meisten vorherigen Studien mit Mischkulturen. In dem anaeroben Festbettreaktor konnte eine stabile Phenolabnahme von 1,48 g/l·d bei einer Phenolkonzentration von 3,7 g/l im Zufluss beobachtet werden. Eine Schockbeladung von um 0,94 g/l erhöhter Phenolkonzentration wirkte sich fatal auf die Stabilität des Reaktors aus. Eine teilweise Wiederherstellung des Reaktors war nur möglich, indem dieser für einen Monat nicht beschickt wurde und die Phenolkonzentration bei einem erneuten Start bei 0,19 g/l gehalten wurde.

Für Untersuchungen der Dechlorierung eines monochlorierten aromatischen Schadstoffes wurde eine Mischprobe des anaeroben Reaktors, die mit kommunalem Klärschlamm behandelt wurde, an 2-CP unter batch Bedingungen angepasst, und anschließend wurde der Überstand benutzt um den kontinuierlichen Festbettreaktor zu beladen. Eine hohe 2-CP-Abbaurate von 0,87 g/l·d in dem anaeroben Festbettreaktor konnte erreicht werden. Die Hemmeffekte, welche von der hohen Chlorphenolbeladung herrührten konnten behoben werden, wenn die Beladung nur kurzzeitig war und vorherige niedrigere Beladungsbedingungen folgten. Die Rate des freigesetzten Chlorids pro Einheit abgebauten 2-CPs zeigte die erfolgreiche Dechlorierung des 2-CPs. Pepton und Hefeextrakt in der Beschickung konnte durch NH_4NO_3 und einem definierten Volumen einer Vitaminlösung ohne eine negative Auswirkung auf die mikrobielle Flora oder den Reaktorbetrieb ersetzt werden. Somit ist der kontinuierliche biologische Abbau von schädlichen aromatischen Anteilen in konzentriertem Abwasser möglich, wenn die mikrobielle Flora in Festbettsystemen an den Schadstoff angepasst ist, was optimal ist, um die angepasste mikrobielle Flora für lange Zeiträume zu halten und die unerwünschte Auswaschung der Mikrobiologie zu minimieren.

Abbreviations

μ	Specific growth rate (h)
μ_{max}	Maximum specific growth rate (h)
2-CP	2-chlorophenol
AFBR	Anaerobic fixed bed reactor
COD	Chemical oxygen demand (g/ l)
d	Days
FID	Flame ionisation detector
g/l	Grams per litre
GC	Gas chromatograph
h	Hours
HAIB	Horizontal flow anaerobic immobilized biomass reactor
HRT	Hydraulic retention time (d)
K_i	Inhibition constant (mmol/ l)
K_m	Michaelis constant (mmol/ l)
K_s	Half saturation constant (mmol/ l)
l	Litre
mg/l	Milligrams per litre
mmol/l	Millimoles per litre
NOE	Nitrogen oxygen equivalents
OLR	Organic loading rate (g COD/ l·d)
ppb	Parts per billion
ppm	Parts per million
S	Substrate concentration (mmol/ l)
SAC_{254}	Spectral absorption coefficient measured at 254 nm
SBR	Sequencing batch reactor
SMBR	Silicone membrane bioreactor
SWW	Synthetic wastewater
TCD	Thermal conductivity detector
TS	Total solids (g/ l)
UAB	Up-flow anaerobic reactor
UASB	Up-flow anaerobic sludge blanket reactor
V	Velocity of the reaction (mmol/ l·h)
V_{max}	Maximum velocity of the enzyme catalyzed reaction (mmol/ l·h)
VSS	Volatile suspended solids (g/ l)
X	Microorganism concentration expressed as a function of optical density at 578 nm
Y	Yield factor (OD biomass/ mmol phenol/ l)

List of Figures

List of Tables

Table of Contents

Chapter 1

Introduction

1.1 Phenol

1.1.1 Background on phenol

Phenol (C_6H_5OH) is an aromatic compound derived from benzene (the simplest aromatic hydrocarbon), by adding a hydroxyl group to a carbon to replace a hydrogen. Phenol is both a man-made chemical and produced naturally. It is found in nature in some foods, in human and animal wastes and decomposing organic material. Phenol is one of the most important class of raw materials in chemical industry, and a variety of compounds are derived from phenol like resins, dyes and paints, pharmaceuticals, etc. The largest single use of phenol is as an intermediate in the production of phenolic resins. It is also used in the production of caprolactam (which is used in the manufacture of nylon 6 and other synthetic fibres) and bisphenol A (which is used in the manufacture of epoxy and other resins). Other uses of phenol are as a slimicide (a chemical toxic to bacteria and fungi characteristic of aqueous slimes), as a disinfectant, and in medicinal preparations such as over-the-counter treatments for sore throats (ATSDR, 2006).

In 2004, the total annual capacity of phenol production approached 6.6 billion pounds and overall production is expected to grow 6 % per year (CMR, 2005). In 1999, almost 90 % of phenol demand was in North America, West Europe and northeast Asia, since these parts of world contain most developed nations (CMAI, 2001). Industrial production of phenol generally takes place via the cumene process. The production of cumene is an indicator of worldwide phenol production (Fig. 1.1). Phenol ranks in the top 50 chemicals, concerning production volumes of chemicals produced in the United States (ATSDR, 1998). New applications in DVD, water bottles, computer parts (i.e. I Macs) and auto-glazing may have even stronger impacts on the growth rate of phenol production. Following small, single releases, phenol is rapidly removed from the air; generally, half is removed in less than 1 day. It is also relatively short-living in the soil (generally, complete removal in 2-5 days). However, it can remain in water a week or more. Phenol can remain in the air, soil, and water for much longer periods of time if a large amount of it is released at one time, or if it is constantly released to the environment. Levels of phenol above those found naturally in the environment are usually found in surface waters and surrounding air contaminated by phenol released from industrial activity and from the commercial use of products containing phenol (ATSDR, 2006).

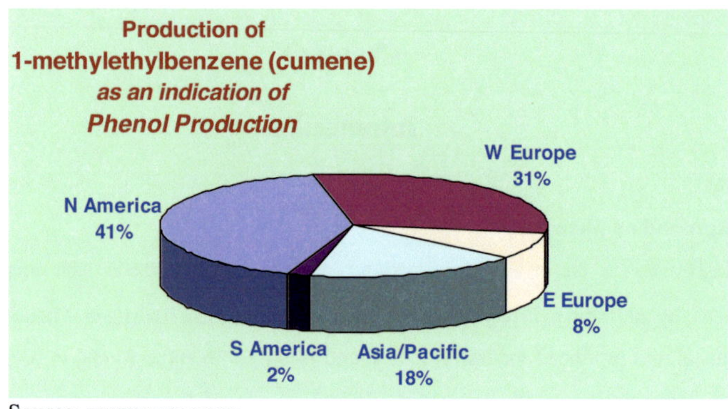

Source: www.uyseg.org

Fig. 1.1 - Worldwide phenol production in 2001

1.1.2 Industrial discharges and toxicity of phenol

Phenol lists among the priority pollutants as it is a highly toxic and bactericidal substance. United States environmental protection agency (USEPA) has prescribed 0.5 mg/l as a 24 hour average, never to exceed 3.4 mg/l, as the limits for industrial discharge of phenol. Whereas the central pollution control board of India (CPCB) has prescribed maximum permissible limit of all phenolic compounds in small scale industrial discharges as 5 mg/l. The concentration of phenol in the industrial wastewaters (Table 1.1) may range from 1 to15,000 mg/l. Therefore treatment of such wastewaters is obligatory. The main problem in treating the phenol containing waste water is the toxicity it exerts to the microbial flora in biological treatment plants. Because of this toxicity there is a danger of partial or complete treatment plant failure, when the microbial flora is not adapted to phenol concentrations in the effluent.

The general bactericidal effect of phenol is due to the disruption of lipid membrane functionality, which is the only barrier between the bacterial cytoplasm and the outer system. Upon exposure of bacteria to bactericidal concentration of phenol i.e. 2 g/l, the lipid membrane of the cell becomes dysfunctional causing leakage of K^+ ions, ATP and eventually nucleotides (Heipieper et al., 1990; 1991). Changes in membrane structure, in response to the phenol have also been observed (Keweloh et al., 1990). Under such conditions both the cytoplasmic and outer membranes showed higher protein to lipid ratios. The higher protein fraction in the membrane makes the fatty acid chains of the lipids more rigid and hence makes the lipoprotein membrane less permeable. The change in membrane structure could be a result of the inhibition of lipid synthesis or an adaptive effect to protect osmotic integrity. Heipieper et al. (1990) have also reported an increased ratio of saturated to unsaturated fatty acids in the cell membrane at bacteriostatic concentrations of phenol.

But bacteria can adapt themselves to the adverse conditions by developing the mechanisms to restore their cell gradients.

Table 1.1: Concentration of phenol in different industrial wastewaters

(Sittig et al., 1975; Kumaran and Parachuri, 1997; Weber et al., 1992; Pinto et al., 2003)

Industrial source	Phenol concentration (mg/l)
Coke Ovens	
Weak ammonia liquor without dephenolization	600-12,000
Wash oil still wastes	430-150
Oil Refineries	
Low-temperature carbonization effluent	3395
Sour water	80-195
General waste stream	10-100
Post- stripping	80
General (catalytic cracker)	40-50
Mineral oil wastewater	100
API separator effluent	1-7
Petrochemical	
General wastewater	50-600
Benzene refineries	210
Nitrogen works	250
Tar distilling plants	300
Aircraft maintenance	200-400
Herbicide manufacturing	210
Other	
Olive oil mill	1,000-1,500
Rubber reclamation	3-10
Orlon manufacturing	100-150
Plastics factory	600-2,000
Fiberboard factory	150
Phenolic resin production	15,000
Dephenolization liquor	3,000
Stocking production	6,000
Fiberglass manufacturing	40-400
Wood carbonizing	500

If the bacteria are adapted to a sub-lethal concentration of phenol, they could reabsorb the K^+ ions at phenol concentrations up to which they were adapted, when cells were supplied with an energy source such as glucose unlike the non adapted cells which tend to loose the ions irreversibly (Heipieper et al., 1990; 1991). The resistance mechanisms consist of isomerisation of *cis*-unsaturated fatty acids to the *trans*-configuration after exposure to phenol. Because the chains of *trans*-fatty acids molecules can align closer together in a biological membrane than those in the *cis*-configuration, a more rigid membrane is formed. Thus cells adapted to phenol by growing on phenol as a carbon source could maintain growth under phenol concentrations of up to 750 mg/l, whereas non adapted *P. putida* P8 cells could not grow at phenol concentrations higher than 250 mg/l (Heipieper et al., 1992). Another mechanism is to increase the ratio of saturated fatty acids to unsaturated acids in the membrane. Keweloh et al. (1991) reported that, *Escherichia coli* K-12 was able to maintain its growth at phenol concentrations up to 1000 mg/l by converting unsaturated fatty acids to saturated ones. Like *trans*-fatty acids, the chains of saturated fatty acids can align together closer in the membrane, which may compensate for the increased membrane fluidity induced by phenol (Schie and Young, 2000). Therefore biodegradation of phenolic wastewaters under aerobic or anaerobic conditions is feasible, once microbial population is adapted to such concentrations.

1.1.3 Mechanisms of phenol degradation

1.1.3.1 Aerobic biodegradation of phenol

After the diffusion of phenol into bacterial cells, molecular oxygen is used by the enzyme phenol hydroxylase to add a second hydroxyl group in *ortho-* position to the one already present. The reaction requires a reduced pyridine nucleotide ($NADH_2$). The resulting catechol (1,2-dihydroxybenzene) intermediate is then degraded via one of two pathways (Fig. 1.2). In the *ortho-* or *β*-ketoadipate pathway, the aromatic ring is cleaved between the catechol hydroxyls by a catechol 1,2-dioxygenase (intradiol fission) (Harwood and Parales, 1996; Stanier and Ornston, 1973). The resulting *cis,cis-* muconate is further metabolized, via *β*-ketoadipate, to carbon dioxide as well as succinate and acetyl-CoA, both of which can be further degraded in the TCA cycle.

In the *meta*-pathway, ring fission occurs adjacent to the two hydroxyl groups of catechol (extradiol fission) (Bayly and Barbour, 1984). The enzyme catechol 2,3-dioxygenase transforms catechol to 2-hydroxymuconic semialdehyde. This compound is metabolized further to form acetaldehyde and pyruvate in the Krebs cycle. The *meta*-pathway appears to allow for the metabolism of a variety of substituted phenolic compounds. Whereas, the *ortho*-pathway is more specific and is used in the catabolism of catechol and precursors of catechol (Feist and Hegeman,

1969). The *ortho*-pathway is more common. Stainer et al. (1966) found that only 8 out of 41 strains of *Pseudomonas putida* were capable of the *meta-* cleavage pathway.

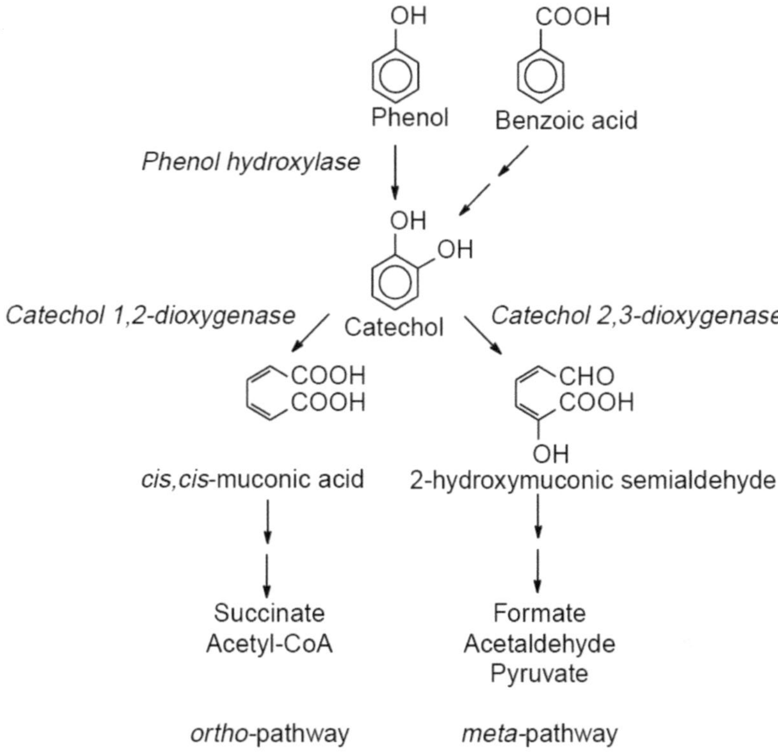

Fig. 1.2 - Metabolic pathways of aerobic phenol degradation
(Modified from Lipthay et al., 1999)

Phenol can be aerobically degraded by a wide variety of microorganisms, including pure bacterial cultures (Table 1.2). Marrot et al. (2006) reported successful treatment of phenol concentrations up to 1 g/l in an immersed membrane bioreactor. They had used activated sludge from a wastewater treatment plant, grown in a continuous culture using phenol as the limiting substrate. However by using a pure strain of *Pseudomonas putida* ATCC 17484 in a fluidised bed (immobilized cultures), > 90 % removal of 42.5 mmol (4g) /l·d phenol has been observed (Goudar et al., 2000).

Table 1.2: Some studies on aerobic phenol removal

Inoculum	Condition	Phenol degradation rate (g/l·d)	Reference
P. putida (MTCC1194)	Shake flasks	0.148	Kumar et al. (2005)
P. cepacia and *B. brevis*	Shake cultures	0.416 and 0.291	Arutchelvan et al. (2005)
Propioniferax like PG-02 and *Comamona sp.* PG-08	Batch	1.68 (PG-02) 0.547 (PG-08) 3.6 (PG-02 + PG-08)	Jiang et al. (2006)
A. citreus	Immobilized in alginate and agar (Batch mode)	0.47	Karigar et al. (2006)
Mixed culutre of *R. rhodochrous, G. sputi, P. putida*	Shake flasks	0.336 to 1.0	Przybulewska et al. (2006)
Activated sludge	Continuous	0.225	Amor et al. (2005)
Mixed culture	Continuous and draw filled mode	0.082 and 12.65	Tziotus et al. (2005)
Mixed consortium	Silica bead packed bed reactor	0.79	Bertin et al. (2006)
Activated sludge	-	> 90 % removal of 4.0	Goudar et al. (2000)
Activated sludge	1) Membrane bioreactor 2) Batch reactor	1) ND[1] 2) 1.3	Marrot et al. (2006)
Mixed consortium	Packed bed with matrix of polymeric hydrogel beads of calcium alginate (1%) and cross linked *N*-vinyl pyrrolidone	1.6 (batch) 1.62 -3 (continuous)- medium 1 0.4 - 3.8 (continuous)- medium 2	Esparza et al. (2006)

[1]ND: not defined

1.1.3.2 Anaerobic biodegradation of phenol

In anaerobic environment oxygen is not available to promote activation of recalcitrant substrates, such as aromatics. The anaerobic biodegradation of organic matters is a complicated process and various pathways are possible. In general an anaerobic process under methanogenic conditions is most popular for the degradation of organic matter which involves syntrophic interaction of microorganisms and three main steps – hydrolysis/fermentation, acetogenesis and methanogensis, respectively. The detailed scheme could be referred to in the book chapter of Gallert and Winter (1999).

Metabolism in the absence of oxygen is dominated by the nature of the available electron acceptors or hydrogen sink reactions. Successively, nitrate, ferric iron, sulfate, and carbon dioxide serve as the preferred electron acceptors for the denitrifying, iron reducing, sulfate reducing and methanogenic bacteria in nitrate rich, iron rich, marine or organic rich systems, respectively. In addition, acid-forming fermentative bacteria and H_2-producing acetogenic bacteria are a part of many anaerobic communities. Chmielowski et al. (1965), were the first to report phenol degradation under methanogenic conditions and thereafter phenol degradation was also reported to occur under nitrate-reducing (Bakker, 1977), sulfate-reducing (Bak and Widdel, 1986), and iron reducing (Lovley and Lonergan, 1990) conditions.

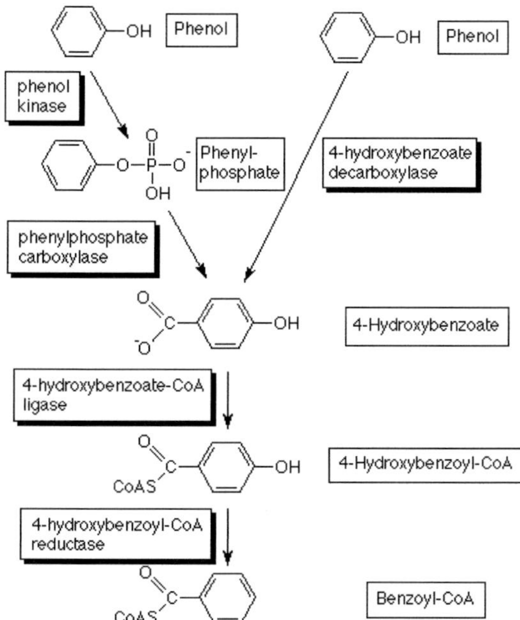

Fig. 1.3 - Metabolic Pathways of anaerobic phenol degradation

(After Lack and Fuchs, 1994; He and Wiegel, 1995)

Knoll and Winter (1987), proposed for the first time a phenol degradation pathway under which phenol was converted to methane via benzoate as an intermediate product. Under denitrifying conditions phenol is first phosphorylated to phenylphosphate and then converted to 4-hydroxybenzoate (Fig. 1.3). Various studies on phenol degradation under different conditions, utilising mixed or pure cultures, have already been published. It has been reported that mixed cultures permit faster phenol degradation than pure cultures (Zache and Rehm, 1989) and systems with immobilized cells are more stable to shock loadings than submerged cultures with free cells (Holladay et al., 1978).

1.2 Chlorophenols

1.2.1 Background on Chlorophenols

Chlorophenols are a group of chemicals in which chlorines (between one and five) have been substituted by hydrogen atom. There are five basic types of chlorophenols: monochlorophenols, dichlorophenols, trichlorophenols, tetrachlorophenols and pentachlorophenols. In all, there are 19 different chlorophenols, formed by replacing from one to five of the non-hydroxyl hydrogens of the phenol molecule with chlorine atoms. These include three monochlorophenols (MCPs), six dichlorophenols (DCPs), six trichlorophenols (TCPs), three tetrachlorophenols (TTCPs) and one pentachlorophenol (PCP). All of the chlorophenols are solids, except 2-chlorophenol, which is a liquid at room temperature. 2-Chlorophenol (2-CP) is one of the three MCPs. Others are 3- and 4-chlorophenol depending upon the position of chlorine substitute at the phenol ring i.e. at *ortho-*, *meta-* or *para*-positions.

The chlorophenols have a strong medicinal taste and odour; small amounts at parts per billion (ppb) to parts per million (ppm), concentrations can be tasted in water. Chlorophenols with at least two chlorines either have been used directly as pesticides or converted into pesticides. Also, chlorophenols, especially 4-chlorophenol, have been used as antiseptic agents. In addition to being produced commercially, small amounts of some chlorophenols, especially the mono- and di-chlorophenols, may be produced when waste water or drinking water is disinfected with chlorine, if certain contaminants are present in the raw water (ATSDR, 1999). In the paper manufacturing, chlorophenols are produced as by products during the bleaching of wood pulp with chlorine (Annachhatre and Gheewala, 1996). Chlorophenols also appear as degradation products of other chlorinated xenobiotics (Bollag et al., 1986).

Chlorophenols can enter the environment while they are being made or used as pesticides. Industrial manufacture of chlorinated phenols proceeds by chlorination of phenol. The annual industrial production of chlorophenols was estimated at 0.2×10^6 metric tones in 1989 (WHO,

1989). During the release of chlorophenols into the environment, most of the contaminants go into water, with very little entering the atmosphere. More volatile compounds like mono- and di-chlorophenols are also likely to go into the air.

1.2.2 Industrial discharges and toxicity of chlorophenols

Various chlorophenols, including 2-CP, are termed as priority pollutants (Keith and Telliard, 1979). Due to their high toxicity, recalcitrance, bioaccumulation, strong odour emission, persistence in the environment and suspected carcinogenity and mutagenity, chlorophenols pose serious ecological problems as environmental pollutants (Armenante et al., 1999; Xiangchun, 2003). No particular information is available for the concentrations of chlorophenol present in the industrial discharges. However the European Community Law has prescribed 50 μg/l (0.39 μmol/l) as the annual mean limit for 2-CP discharges into coastal and surface waters (Graham, 2002).

The recalcitrance of chlorophenols results from the carbon–halogen bond, which is cleaved with great difficulty and from the stability of their aromatic structure, resulting in their accumulation in nature (Farrell and Quilty, 2002). The mechanism of bacterial toxicity is not known but it is assumed to be similar as for the phenols (Sec. 1.1.2). Microoragnisms present in municipal sewage sludge that have not been previously exposed to chlorinated phenols have been shown to degrade 50 μg/l (0.39 mmol/l) *ortho*- and *meta*-chlorophenols in 6 weeks (Boyd and Shelton, 1984).

Treatment of industrial wastewater with biological methods has attracted more attention than mechanical and chemical methods, e.g. adsorption or wet oxidation. Biological treatment of industrial wastewater that contains high concentrations of chlorophenols is difficult. Conventional treatment systems both, aerobic or anaerobic, often fail to achieve a satisfactory efficiency in treating high concentrations of chlorophenol-containing wastewater due to its toxicity or inhibitory effect on microorganisms, unless these were adapted to or selected for degradation of such contaminants. However many types of microorganisms, such as *Acinetobacter sp.*, *Desulfomonile tiedjei*, *Desulfitobacterium chlororespirans or Alcaligenes sp.* (Kim and Hao, 1999; Holliger et al., 1999; Gallego et al., 2001) are known to utilize chlorophenols as their sole carbon and energy source. For large scale industrial wastewater treatment, mixed consortia rather than pure cultures are preferred, as they are easier to handle and contain diverse microbial species capable of withstanding various unfavourable operation fluctuations. Research has been carried out on the biological treatment of chlorinated phenols in wastewater, groundwater, soil and compost under aerobic, anaerobic or combined processes. The degradation of highly chlorinated phenols is often incomplete and results in the accumulation of monochlorophenols, e.g. of 2-chlorophenol in the system. Therefore removal of accumulated monochlorophenols is very important for the complete mineralisation.

1.2.3. Mechanisms of 2-chlorophenol (2-CP) degradation

1.2.3.1 Aerobic biodegradation of 2-CP

The aerobic degradation of chlorophenols occurs through the formation of corresponding catechols (Spain and Gibson, 1988). In the aerobic degradation pathway, lower chlorinated phenols (mono- and di-chlorines) are initially attacked by monooxygenases yielding chlorocatechols as the first intermediates (chlorocatechol pathway), which are subject to ring cleavage prior to dechlorination. The higher chlorinated phenols (3 to 5 chlorines) are converted to chlorohydroquinones as the initial intermediates (hydroquinone pathway). Subsequent reactions progressively remove chlorines from the ring prior to ring cleavage. In most strains studied, the conversion of chlorocatechols generally proceeds via *ortho*-cleavage. The alternative pathway via *meta*-cleavage is less common because the 2,3-dioxygenase is inactivated by 3-chlorocatechols (3-CC) and the cleavage product from 4-CC, 5-chloro-2-oxymuconic semialdehyde, is toxic to bacteria (Solyanikova and Golovleva, 2004).

Beside above said, several other pathways of chlorophenol degradation by aerobic bacteria could be present, e.g. Nordin et al. (2005) observed 4-CP dechlorination by *Arthrobacter* via 1,4-hydroquinone as the intermediate. Chlorinated phenols can serve as sole carbon and energy sources supporting growth (Rutgers et al., 1997; Hollender et al., 2000; Yang et al., 2005) or can be degraded as a co-substrate (Cobos-Vasconcelos et al., 2006; Loh and Wu, 2006). Most of the aerobic chlorophenol-degrading strains belong to the genera, *Mycobacterium* and *Sphingomonas* (Field and Alvarez, 2007). Beside bacteria, several white-rot fungi species, e.g. *Phanerochaete chrysosporium* (Yum and Peirce, 1998), *Trametes versicolor* (Pallerla and Chambers, 1998) are also able to degrade chlorophenol.

1.2.3.2 Anaerobic biodegradation of 2-CP

Chlorinated phenols are readily metabolized by bacteria under anaerobic conditions. Anaerobic degradation of chlorophenols is initiated by reductive dechlorination, in which chlorine substituent are removed with concurrent addition of hydrogen atoms. Reductive dechlorination generally requires the input of electron donating substrates (Fig. 1.4). It is difficult to determine, whether the reductive dechlorination is co-metabolic or beneficial to microorganisms. In some cases, reductive dechlorination is known to be linked to growth due to the use of the chlorinated phenols as electron acceptors (halorespiration). Also, there are a few examples in which low chlorinated phenols are clearly the carbon and energy source (the electron donor) to microorganisms (Field and Alvarez, 2007). In some cases under denitrifying conditions, chlorophenols were converted to 3-chlorobenzoate (Fig. 1.4), which was completely mineralized by a number of pure culture isolates of the genera *Thauera*, *Pseudomonas*, and *Ochrobacterium* and thus these bacteria used this chlorinated compound as a sole source of carbon and energy (Song et al., 2000). The majority of the chlorophenol halorespiring bacterial isolates belong to the genus *Desulfitobacterium* (Villemur et al., 2006). No pure cultures, however, have so far been obtained that are able to oxidatively dehalogenate haloorganic compounds under sulfidogenic, or Fe(III)-and Mn(IV)-reducing conditions, although complete mineralization of a variety of organohalides under these conditions has been repeatedly observed for mixed cultures (Bradley, 2003; Löffler et al., 2003). Cell yields observed during halorespiration are relatively low but remarkably high specific activities are noted for halorespiration of chlorophenols. The combination of relatively high conversion rates and low cell yields accounts for the high specific activities that far exceed those of aerobic chlorophenol-degrading bacteria.

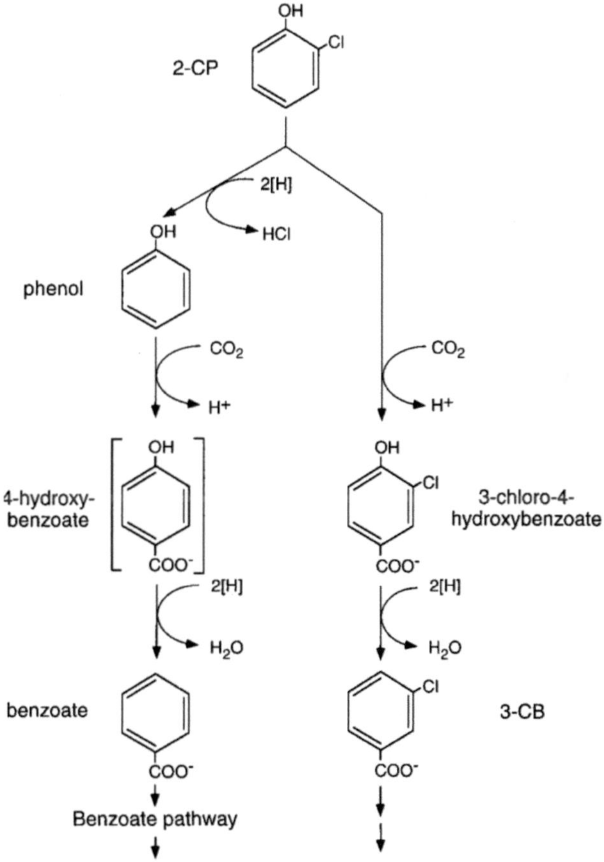

Fig. 1.4 - Pathways of anaerobic 2-CP biodegradation (from Becker et al., 1999)

The degradation of 3-CP or a mixture of 2-CP, 3-CP and 4-CP as sole sources of carbon and energy was evaluated in hybrid UASB anaerobic filter reactors (Krumme and Boyd, 1988). The reactors could accommodate a volumetric loading of 20 g (155.5 mmol/l) CP per m^3 reactor per day with greater than 90 % removal of chlorophenols. In an another study, sulfate-reducing consortia enriched from estuarine sediment were maintained on either 2-CP, 3-CP, or 4-CP as the only source of carbon and energy for over 5 years (Haggblom and Young, 1995). Kazumi et al. (1995), recovered Fe^{2+} from Fe^{3+} reduction during chlorophenol degradation under iron reducing conditions. Only at the beginning of this decade, chlorophenol degradation linked to denitrification has been demonstrated with an enrichment culture utilizing 2-CP derived from activated sludge (Bae et al., 2002). An enrichment culture from sewage sludge was able to reductively dechlorinate 2-CP and 2,6-DCP to phenol and HCl in medium containing yeast extract and peptone (Dietrich and Winter, 1990). Methanogenic enrichment cultures developed from municipal digester sludge mineralized radiolabeled [^{14}C] 4-CP, [^{14}C]2-CP and [^{14}C] 2,4-DCP by > 90 % to $^{14}CH_4$ and $^{14}CO_2$ (Boyd and Shelton, 1984). The results taken as a whole suggest that chlorophenols utilized as a carbon and energy source are first reductively dechlorinated to phenol. Subsequent phenol mineralization provides the energy and carbon to support growth as well as electrons to support the initial dechlorination.

1.3 Objectives of the study

Because of their wide applications and known toxicity on microbial flora, phenol and 2-chlorophenol are model toxic substances for biodegradation studies. 2-CP is a common intermediate of biodegradation of highly chlorinated phenols. It very likely accumulates and thus affects wastewater treatment processes. A lot of biodegradation studies with these compounds are already available; still experimental data on phenol and 2-chlorophenol removal at high concentrations are scarce. Most of the studies commonly deal with only 5 to 15 mmol phenol/l (470 to 1417 mg/l) or 0.1 to < 2 mmol 2-chlorophenol/l (12.8 to < 257 mg/l) in wastewater. The main questions addressed in this thesis are, how much phenol and 2-chlorophenol can be tolerated and degraded by microorganisms under continuous operating conditions for long time periods? And how suspended and fixed bed systems respond to transient loading and operation.

Degradation studies were first carried out in aerobic, anaerobic and anoxic suspended bed reactors and then in fixed bed reactors in order to find out effective methods for treating the phenol containing wastewaters and to understand the dechlorination of a monochlorinated aromatic pollutant by mixed consortium. The research approach consisted of obtaining suitable seed sludge through acclimatisation and then increasing the phenol and 2-CP concentration gradually in the reactor influent, unlike the usual strategy, where the concentration of toxic compound in the influent is kept low and the loading is increased by decreasing the HRT. The system response was monitored for shock loadings by measuring the bulk phase concentrations. These observations would provide useful information for industrial wastewater treatment of effluents containing high and fluctuating concentrations. The data obtained from the best system was used to calculate the biokinetics parameters of phenol degradation. Since most of the industrial wastewater treatment processes deal with sludges that consist of mixed cultures only, the knowledge of microbial growth, substrate inhibition and utilization is useful to predict the fate of a pollutant during treatment processes. The main objective of this study were

1. To determine the treatment capacity of suspended and fixed biofilm reactors and to demonstrate the ability of microbial consortia of domestic sewage sludge to adapt to high concentrations of the toxic pollutants such as phenol (up to 3764 to 4705 mg/l) and 2-chlorophenol (up to 2600 mg/l), respectively. The goal was to harness the microbial ability of adaptation to toxic substances without the need of any pre-treatment of highly concentrated wastewater.
2. To investigate the changes after the shock loading of the phenol applied to the stable system and possible recovery of the reactor.
3. To find out maximum degradation rates of phenol and 2-chlorophenol.

Chapter 2

Materials and methods

2.1 Analytic

The samples were tested for the various parameters as required. All the chemicals and gases used for testing were of analytical grade, purchased from Carl Roth (Karlsruhe), Merck/VWR (Darmstadt), Fluka (Taufkirchen) or Sigma (München). The gases used for chromatographic analysis were of grade 5.0 (Linde AG, Pullach). For all photometric analysis a UV LKB Biochrom Ultrospec II spectrophotometer was used. The centrifuges used during the study period were Hermle Z 233 M-2, Labfuge 15000 or Eppendorf 5403, Germany. The mechanical shaker used during batch studies was Lab-shaker model Kühner from B. Braun, Germany.

2.1.1 Chemical Oxygen Demand (COD)

The chemical oxygen demand represents the oxygen equivalent of the organic matter content of a sample that can be fully oxidised to carbon dioxide with a strong oxidising agent under acidic conditions. The method of Wolf and Nordmann (1977) was used for COD determinations with potassium dichromate ($K_2Cr_2O_7$) as the oxidising agent in a solution of H_2SO_4 and H_3PO_4 acids (3:1). The catalyst was Ag_2SO_4. The sample was centrifuged and 1 ml of respectively diluted supernatant was mixed thoroughly with 1.5 ml of COD reagent containing the oxidizing agent and refluxed for 2 hours on a thermoblock (Thermo, Bielefeld) at $150\,^{0}C$. After cooling, the absorbance of the built green colour of released $Cr\ (III)^{+}$ due to $K_2Cr_2O_7$ reduction was determined with a spectrophotometer at 615 nm. The concentration of the unknown samples was calculated by comparison with a standard curve prepared with potassium hydrogen phthalate ($C_8H_5KO_4$) (200-1400 mg/l COD).

2.1.2 Phenol and 2-chlorophenol (2-CP)

The phenol and 2-CP concentrations in the samples were determined by gas chromatography (CP 9001) employing a flame ionisation detector (FID) with a hydrogen/air flame. The samples were centrifuged at 12,500-15,000 rpm for 5-6 minutes and the supernatant was analysed immediately or frozen for later analysis. The sample volume used for analyses was 1 μl, injected with a 10 μl Microliter$^{®}$ TM # 701 syringe (Hamilton Bonaduz, Switzerland) into the GC. Nitrogen served as the carrier gas. A VF-Xms WCOT fused silica column (Varian) at an increasing oven temperature of $105\,^{0}C$ to $145\,^{0}C$ with a rise of $10\,^{0}C$/min at an injector temperature of $275\,^{0}C$ and a detector temperature of 250 ^{0}C was used for separation. The flow settings for the gases were: hydrogen, 30 ml/min; nitrogen, 30 ml/min and air, 250 ml/min. The carrier gas pressure at the flow control panel was 100 kPa. The peaks for phenol and 2-CP were analysed and quantified with Maestro Chromatography Data System, version 2.3.

2.1.3 Biogas composition

Daily biogas production from the anaerobic reactors was measured with a Gas meter (Ritter, Bochum-Langendreer, Germany), which was based on the principle of water displacement. The composition of biogas was analysed by injecting 0.1 ml of sample with a 0.5 ml Pressure-LoK$^®$ syringe (VICI Precision sampling, USA) into a gas chromatograph (Chrompack CP 9001) equipped with a micro volume Thermal Conductivity Detector (TCD). A biogas standard consisting of 60 % CH_4 and 40 % CO_2 (or when required also prepared with desired content of oxygen or/and hydrogen), was injected first as the reference concentrations for the GC calibration. Nitrogen was the carrier gas at a flow rate of 30 ml/min with the following temperature settings: column 110 ^{0}C, detector 220 ^{0}C and injector 250 ^{0}C, respectively. The filament temperature of the TCD was set automatically approximately 100 ^{0}C above the detector block temperature. The pressure at the control panel for both analysis and reference regulators was 160 kPa. A capillary column CarboPlot$^®$ P7 WLD FS 25 x 0.53 m (Chrompack, Netherlands) served for separation.

2.1.4 Fatty acids

The concentration of fatty acids was determined by gas chromatography (United Technologies Packard 437A) with FID detection using a Teflon column packed with Chromosorb 101 (Sigma, München). The samples were centrifuged for 5-6 minutes at 12,500-15,000 rpm and the clear supernatant was acidified 1: 1 with 4 % phosphoric acid (H_3PO_4). A 10 μl syringe (TM # 701, Hamilton Bonaduz, Switzerland) was used to inject a 1 μl sample into the injection port of the GC. A standard of mixed fatty acids, i.e. acetate, propionate, i- and n-butyrate and i- and n-valerate (5 mmol/l each), were injected as reference before the analysis of the test samples. Nitrogen was the carrier gas at an oven temperature of 180 ^{0}C and an injector and detector temperature of 210 ^{0}C each. The gas flows for N_2 and H_2 were 30 ml/min each and the flow rate for air was 250 ml/min. A Teflon column packed with Chromosorb 101 (Sigma, München) served for separation of fatty acids.

2.1.5 Chloride (Cl$^-$)

The concentration of the Cl$^-$ ions was determined in the test samples by ion chromatography (Dionex ICS-90) employing a AS9-HC 4 x 250 mm (IonPac$^®$) analytical column and using 9 mM

Na_2CO_3 as an eluent at a flow rate of 1 ml/min and 2 N H_2SO_4 acid as a regenerent. The detection was done with suppressed conductivity (Autosuppression® Recycle mode) with initial conductivity of 170-177 μS. For analyses, the centrifuged samples were filtered through 0.45 μm cellulose acetate membrane filter (Whatman OE 67) and injected on port with a 1 ml syringe (Terumo, Belgium). A Cl^- standard curve (0.5 mg/l to 11 mg/l) was used for calibration of the instrument and the concentration of unknown samples was calculated automatically by Chromeleon ® software.

2.1.6 Nitrate-N (NO_3^--N)

The concentration of NO_3^--N was determined using the 2,6-dimethyphenol (DMP) method of German Standard Methods DEV D9 (DIN 38 405). The 2,6-dimethyphenol reagent (1 ml) was added to centrifuged sample (1 ml) acidified with 8 ml of a 1:1 mixture of H_2SO_4 (95-97 %) and *ortho*-H_3PO_4 acid. A colour development after 15-30 minute incubation due to the NO_3 and DMP interaction was measured with a spectrophotometer at 324 nm. A standard NO_3-N curve (0-25 mg/l) was used as a reference for calculation of unknown sample concentrations. Since NO_2 causes an interference in the analysis, it was removed by adding some crystals of Amidosulfonic acid before acidification of the sample.

2.1.7 Nitrite-N (NO_2^--N)

The presence of nitrite (0-80 mg/l) in the samples was determined with Merckoquant® NO_2^- test strips (Merck, Darmstadt) and measurement in a spectrophotometer according to German Standard Methods DEV D10 (DIN EN 26777). Nitrite builds up an intensive pink colour in acidified solution (pH 1.9) with sulfanilamide and N-(1-Naphtyl)-ethylendimine, which could be measured at 540 nm. A standard curve was used for determining the unknown sample concentrations. For spectrophotometeric analyses, 50 ml of sample mixed with 1 ml of the nitrite reagent was incubated for 45-90 minutes for colour development.

2.1.8 Total solids (TS) and volatile suspended solids (VSS)/biomass

TS and VSS were determined according to Standard Methods of Wastewater Analysis (method 2540, APHA, 1992). To calculate TS, a defined volume of homogenised sample was kept in a hot air oven (Memmert, Germany) at 105 ^{0}C for evaporation of moisture till a constant weight was obtained. For determining VSS or the organic fraction, the sample after TS determination was heated in a Muffle Furnace (Heraeus Instruments, Hanau) for 2 hours at 550 ^{0}C and the ash content was subtracted from the weight before oxidation.

The growth of biomass was also measured by spectrophotometer through the absorbance inferred due to the turbidity caused by biomass at 578 nm.

2.1.9 Spectral absorbance co-efficient (SAC)

The SAC was determined according to Sontheimer et al., 1986. The absorbance of sample at 274 nm was correlated with C=C- und C=O- double bonds and conjugated double bond systems present in aromatic compounds.

2.1.10 Glucose

Glucose was analysed according to the method of Miller (1959). The sample (0.25 ml) was mixed with 0.25 ml deionised H_2O and 1.5 ml dinitosalicylic acid reagent and heated at 100 ^0C for 5 minutes in a thermoblock (Liebisch, Germany). After cooling and addition of 3 ml deionised H_2O, the colour intensity was measured at 550 nm with a spectrophotometer. A glucose standard curve (0-1000 mg/l) was used as a reference.

2.1.11 Microscopy

The wastewater and sludge samples were examined by light and phase contrast microscopy using a Zeiss Axioscope microscope with camera (Axiocam), connected to a computer for picture analysis with Axioscope software. Presence of methanogenic bacteria in the samples was determined by autofluorescence using fluorescence microscopy with UV light (420 nm). A hydrogen-transferring coenzyme F_{420} emits a blue-green fluorescence colour UV light. This coenzyme is a deazaflavin and is characteristic of methanogenic bacteria and some *Streptomyces species*.

2.1.12 pH

The pH was analysed using a WTW InoLab Mutlilevel 1 pH-meter, Weilheim. The electrode used was WTW SenTix 61.

2.2 Experimental

Both continuous and batch studies were carried out for biodegradation of synthetic wastewater. For continuous studies 6 bioreactors (3 suspended bed and 3 fixed bed reactors) were employed. The peristaltic pumps used to pump the feed into the reactors were from Gilson, model Minipuls 3, France. The wastewater recirculation pumps used for the suspended bed reactors were Watson Marlow 604U, model IP55 pumps (Washdown, England).

To ensure stable influent COD values in all the reactors, the influent bottles were refrigerated at 4 ^{0}C and connected to the feed pump via an influent pipe. At each feed refill, throughout the study period the influent tubings were autoclaved and the influent bottles was replaced with sterilized medium bottles, in order to minimise possible biodegradation by contaminates outside the reactor or backward growth of a biofilm at the inner surface of the tubing. At high phenol concentrations (> 30 mmol/l or 2.8 g/l) and in the absence of glucose, non-autoclaved medium from clean carboys was pumped into the reactor through thermally sterilized tubings, because the high influent phenol concentration itself functioned as bactericidal agent for the reservoir milieu. However tubing's were continued to be sterilised in order to reduce the danger of backward growth.

2.2.1 Aerobic suspended bed reactor

A schematic diagram of the aerobic fixed bed reactor is illustrated in Figure 2.1. The reactor consisted of a hard poly vinyl chloride (PVC) plastic body, aeration system, influent and effluent streams and reservoirs. This reactor was an up-flow system. Aeration was provided along with the feed from the bottom. The liquid working volume of the reactor was 1700 ml. The reactor was inoculated with activated sludge taken from the municipal sewage treatment plant, Neureut. The reactor was started with glucose and minerals containing synthetic wastewater (Table 2.1) having a final COD of 5 g/l and maintained at room temperature (20-25 ^{0}C) throughout the operation period. This reactor was operated at various hydraulic retention times (HRT) and organic loading rates (OLRs) (Sec. 3.1.1.1). Phenol was added into the feed solution when the steady state was achieved. Influent and effluent streams of the reactor were monitored for COD, phenol, pH, HRT, OLR, total and volatile suspended solids.

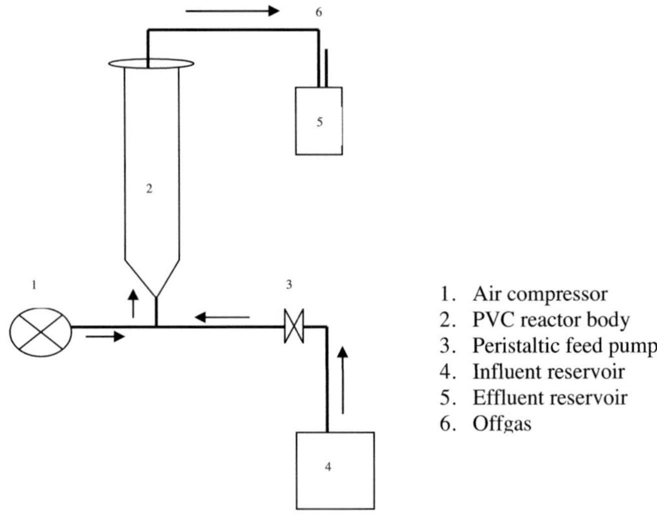

Fig. 2.1 - Schematic diagram of the aerobic suspended bed reactor

2.2.2 Anaerobic suspended bed reactor

A scheme of the reactor design used for the study is shown in Figure 2.2. The primary components of the system were the main reactor compartment, the recycle system, the constant temperature jacket and a thermostat, the influent/effluent reservoir and the influent/effluent streams and a gas outlet connected to a gas meter. The reactor had a glass body fitted with a plastic gas head and a steel bottom containing a sieve. The reactor was completely filled and had a liquid volume of 4500 ml (8 cm internal diameter and 90 cm height). This was an up-flow system. The wastewater for the recirculation was withdrawn slightly below the overflow port and recirculated to the bottom of the reactor. Both, the effluent and the recirculation ports were located at the side of the reactor on the top. The influent pipe was connected to the recirculation pipe at the bottom in order to supply the fresh feed along with the recirculation. This arrangement allowed the complete mixing of reactor liquor and biomass. This reactor was maintained at 37 ^{0}C through the thermostatic water jacket surrounding the reactor body.

The reactor seed was the sludge of an anaerobic bioreactor of the municipal wastewater treatment plant, Berghausen. The reactor was started with glucose and minerals containing synthetic wastewater (Table 2.1) with a final COD of 5 g/l and operated at various loading conditions by changing the HRT (Sec. 3.1.1.2). Phenol was added in the feed solution when a steady state was achieved. The reactor performance was monitored for influent and effluent COD, phenol, gas composition, volatile fatty acids and pH.

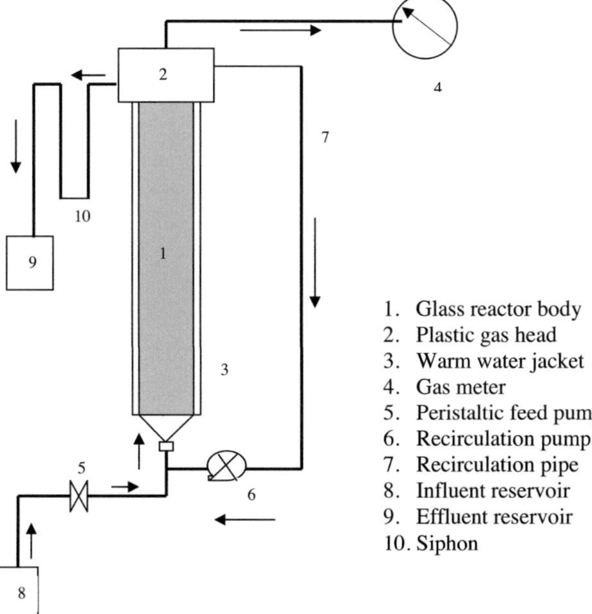

Fig. 2.2 - Schematic diagram of the anaerobic suspended bed reactor

2.2.3 Anoxic suspended bed reactor

The schematic design of the anoxic reactor (Fig. 2.3) consisted of a glass cylinder (5.5 cm internal diameter and 45 cm height) fitted with a recirculation system, influent/effluent reservoir, influent/effluent pipes, and a gas outlet connected with a gas meter (Ritter, Bochum). This reactor was an up-flow system. A thick rubber stopper was provided at the top of the reactor. The effluent and recirculation pipes were drilled through this stopper. The influent pipe was connected to the recirculation pipe at the bottom. The recirculation stream worked in upward direction. The total working volume of the reactor was 1740 ml.

The reactor inoculum was activated sludge taken from the denitrification chambers of the municipal wastewater treatment plant, Neureut. The reactor was started with glucose (as the source of carbon), potassium nitrate (as an electron acceptor in microbial metabolism) and a minerals containing synthetic wastewater (Table 2.1) with a final COD of 5 g/l. Phenol was added to the feed solution when the reactor was operating at steady state conditions. Various organic loading rates were applied to the reactor both in absence and in presence of phenol in the feed. The reactor was monitored for influent and effluent COD, phenol, NO_3-N, NO_2-N, volatile fatty acids and gas composition (measured from time to time).

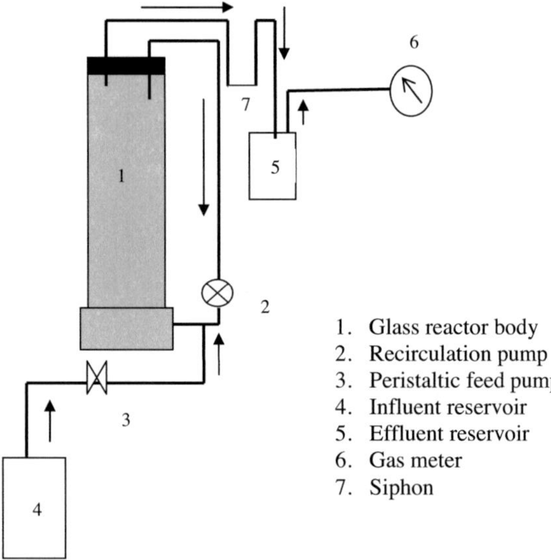

Fig. 2.3 - Schematic diagram of the anoxic suspended bed reactor

2.2.4 Carrier material for the fixed bed reactors

The packing materials for all the fixed bed reactors used in this study was clay beads -Liapor®, Forchheim (11-13 mm) with an average diameter of 12 mm. The carrier material served as a substratum for biofilm growth. Liapor® clay beads (illustrated in Fig. 2.4) are an inert material that is quite resistant to abrasion. This material is cheap and readily available in local garden markets. The adsorption and desorption of phenol from the saturated clay beads were tested and found to be absent. Thus the use of this material did not interfere with biological degradation of organic matter and with analysis in the water phase. Porosity of this material was determined by a simple laboratory experiment involving volume/density measurements $\pm$ vacuum of saturated and unsaturated material and calculation of the pore volume. The wet porosity of this material is 42 % and the dry porosity is 48 %. Before packing the clay beads into the reactor, they were washed thoroughly and soaked in tap water for approximately 40 hours. The density of the material is 273.97 g/l. To avoid floating of the beads and eventual clogging of the influent and effluent pipes, plastic grids were placed at both top and bottom of each fixed bed reactor.

2.2.5 Aerobic fixed bed reactor

The scheme of the aerobic fixed bed reactor used in this study is shown in Figure 2.4. The reactor body consisted of a glass cylinder (5 cm internal diameter and 30 cm height) and primary components of the reactor consisted of influent and effluent pipes and reservoirs, respectively, aeration system and support material for biofilm formation. The feed was provided up-flow along with the aeration. The fixed bed of the reactor was Liapor® clay beads. The empty reactor volume was 500 ml. The reactor was filled with carrier material to ca. 80 % (140 gm clay beads). Thus the working liquid volume of the reactor was 200 ml. An arrangement at the top of the reactor was made for the air outlet. The effluent was collected from the side opening at the top of the bioreactor. The reactor was operated at room temperature ($25 \pm 5\ ^0$C), with a gradual change to simulate the in-situ situation). Effluent from the aerobic suspended bed reactor (Sec. 2.2.1) was used as inoculum for start- up of the reactor. The parameters analysed for this reactor were the same as for the aerobic suspended bed reactor. The influent feed consisted of 0.19 g/l phenol and glucose with the total COD of 5 g/l. The concentration of phenol was gradually increased up to 5.2 g/l, which resulted in changes in influent COD throughout the reactor operation. The reactor was operated at different loading rates, the operational phases of the reactor are discussed in results (Sec. 3.1.2.1.1).

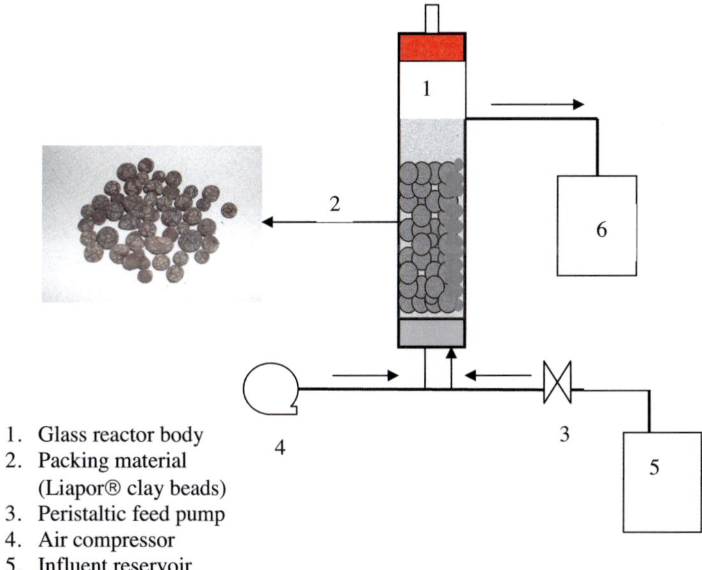

1. Glass reactor body
2. Packing material
    (Liapor® clay beads)
3. Peristaltic feed pump
4. Air compressor
5. Influent reservoir
6. Effluent reservoir

Fig. 2.4 - Schematic diagram of the aerobic fixed bed reactor
(The insert photo shows the carrier material used for the fixed bed)

2.2.6 Anaerobic fixed bed reactor

A scheme of the anaerobic fixed bed reactor is shown in figure 2.5. This anaerobic fixed bed reactor used for the treatment of phenol containing wastewater consisted of a glass cylinder with a working volume of 2000 ml (internal diameter 9 cm and height 54 cm) maintained at 37 ± 2^0C through a jacket of water kept warm through a thermostat. The influent was provided at the bottom and effluent was collected from the upper side of the reactor via a siphon. At the top, a gas outlet was connected to a wet gas meter (Ritter, Bochum). The reactor was packed to ca. 80 % (volume) with Liapor® clay beads as the carrier material for sludge retention/biofilm formation. The reactor was charged with the supernatant of effluent from the anaerobic suspended bed reactor, already in operation for 1.3 years (Sec. 2.2.2). The influent (Table 2.1) of the reactor consisted of glucose as its main substrate and phenol as co-substrate with an initial COD of 5 g/l. The proportions of the glucose and phenol were reversed after bacterial acclimatization as discussed in table 3.6. The changes in the feed and operation at different OLRs and HRTs are discussed in section 3.1.2.2. The reactor was monitored for similar parameters as the anaerobic suspended bed reactor (Sec. 2.2.2).

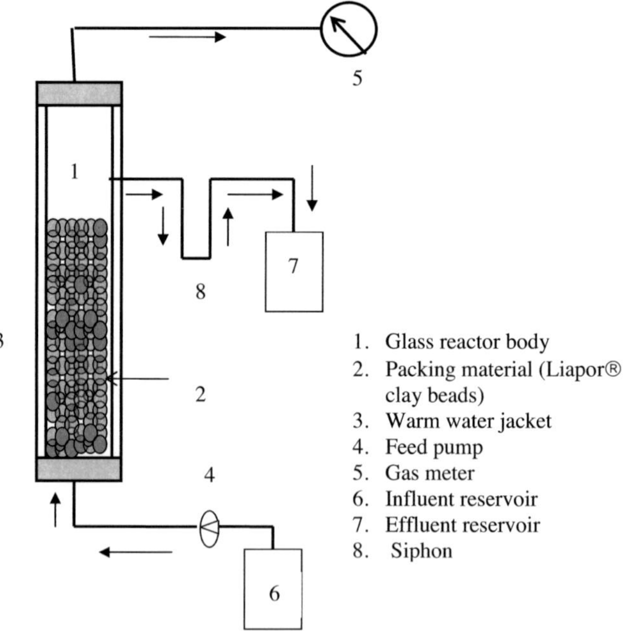

Fig. 2.5 - Schematic diagram of the anaerobic fixed bed reactor for phenol treatment

2.2.7 Preparation of a (2-CP)-degrading seed inoculum

The sludge for inoculum was taken from an anaerobic reactor of the municipal sewage treatment plant, Berghausen. It was incubated at 37 ^{0}C with 50 mg/l of 2-CP and its degradation was monitored. Upon complete removal after about 3 weeks, this sludge was spiked again twice with 2-CP to enrich 2-CP-degrading organisms. Then the turbid supernatant (containing most of the bacteria) was used as a seed for start–up of a 2-CP treating continuous reactor.

2.2.8 Anaerobic fixed bed reactor for 2-CP treatment

For the treatment of 2-CP containing wastewater, an anaerobic fixed bed reactor was started by introducing the supernatant of the previously 2-CP-acclimatised sludge after gravity sedimentation of the solids for 1 h, which served as the seed inoculum (Sec. 2.2.7). The scheme of this reactor was similar to the anaerobic fixed bed reactor treating phenol (Fig 2.5). This reactor consisted of a glass cylinder (9 cm internal diameter and 38 cm height) which was 80 % packed with Liapor$^®$ clay beads. The liquid working volume of the reactor was 1200 ml. The reactor was maintained at 37 ± 2 ^{0}C through a jacket of warm water regulated with a thermostat. The influent was provided at the bottom using a peristaltic pump and the same amount of effluent was displaced at the top of the reactor through a siphon connected to a wet gas meter (Ritter, Bochum). The influent feed was a synthetic wastewater consisting of 2-CP as the main carbon substrate along with yeast extract and peptone (Table 2.1). The initial COD at the start of the reactor was approximately 1000 mg/l. The concentration of 2-CP was increased gradually from 250 to 2600 mg/l throughout the reactor operation; therefore the organic loading rate was also changed accordingly. The operating conditions are presented in Table 3.11 as well as in section 3.3.

2.2.9 Batch assays

2.2.9.1 Aerobic batch assays - Phenol degradation profile

To study the biodegradation of phenol, batch assays along with controls were undertaken in 250 ml Erlenmeyer flasks, closed with cotton plugs. Each flask (in duplicate) contained 50 ml of enrichment medium (Table 2.4) and was incubated on a mechanical shaker at 25 ± 3 ^{0}C and 110 rpm. The flasks contained 10 % sludge as an inoculum, taken from the continuously run reactor. The amount of phenol and other conditions are mentioned along with the results (Sec. 3.2.1.1).

2.2.9.2 Aerobic batch analysis – Substrate co-metabolism with phenol acclimatised sludge

This test was carried out with 3 types of inoculum- a high strength phenol degrading sludge under batch conditions, non-acclimatised sludges and an enrichment culture. The preparations and test was carried out as following. All the assays were in duplicate.

2.2.9.2.1 Preparation of the high strength phenol degrading seed sludge

Effluent from the aerobic fixed bed reactor, degrading phenol at a rate of 10 mmol/l·d (0.94 g/l·d) was used as the seed inoculum and maintained in 500 ml Erlenmeyer flasks closed with cotton plugs and cultured on a mechanical shaker at $25 \pm 3\ ^0C$. The sludge was fed initially with 0.94 g/l of phenol. Phenol dosage was increased gradually with each consecutive feeding. Once in a fortnight, the sludge was centrifuged at 11,000 rpm for 10 min. The supernatant was discarded, and the cell pellet was suspended in phosphate buffer (pH 7.2, 50 mmolar) or in an enrichment media (Table 2.4). The sludge was maintained with 35-40 consecutive phenol feedings from 10 mmol/l to 25 mmol/l. This phenol adapted sludge was used for further experiments.

2.2.9.2.2 Preparation of non-acclimatised sludge

Fresh activated sludge was taken from the municipal sewage treatment plant, Neureut. This sludge was kept on a mechanical shaker in Erlenmeyer flasks for 24 h at $25 \pm 3\ ^0C$. Then the sludge was centrifuged for 10 min at 10,000 rpm. The sludge pellet was suspended in enrichment media (Table 2.4).

2.2.9.2.3 Enrichment culture from adapted sludge

A bacterial culture was enriched from the phenol adapted sludge (Sec. 2.2.6.1.1) by 7 consecutive transfers of 5 % cell suspension into enrichment media (Table 2.4), containing phenol as a sole carbon source.

2.2.9.2.4 Experimental conditions

The experimental process was same as described in section 2.2.6.1 except that the assays with the enrichment culture were monitored for phenol degradation and biomass growth. As soon as the decrease in phenol concentration was observed, glucose (6 mmol/l) was added in assay 6P+6G and acetic acid (6 mmol/l) was added in assay 6P+6A, respectively (Sec. 3.2.1.4).

2.2.9.3 Biodegradation kinetics of aerobic sludge

Effluent from the fixed bed reactor, degrading phenol at a rate of 10 mmol/l·d was used as the seed inoculum in 500 ml Erlenmeyer flasks closed with cotton plugs and maintained on a shaker at 25 $\pm$

3 ^{0}C. A bacterial culture was enriched from the phenol adapted sludge by taking 10 % cell suspension in a mineral medium (Table 2.2). For studies of biodegradation kinetics, 50 ml of the above said cell suspension in mineral medium was used in 250 ml cotton-plugged Erlenmeyer flasks (duplicate). The assays were kept on a shaker at a temperature of 25 $\pm$ 2 ^{0}C. To all the assays, desired concentrations of phenol from 0.25 mmol/l to 7 mmol/l were added. The assays were monitored for phenol degradation and biomass growth (Sec. 3.2.1.5).

2.2.9.4 Anaerobic and anoxic batch assays

Anaerobic and anoxic batch assays were carried out in serum bottles sealed with rubber stoppers and aluminium caps. The volume of each bottle was determined by weighing empty and fill weights of the bottles with the respective stoppers and deducting the exact volume. All bottles had different volumes which ranged from 116-119 ml. The inoculum was either suspended sludge (from anaerobic and anoxic reactors) or effluent from the anaerobic fixed bed reactors (phenol or 2-CP treating reactors). The total working volume was kept at 50 ml in each assay. The required amount of substrates (in form of concentrated solutions) in each assay was added as mentioned along the results. The anaerobic conditions in the assays were maintained by evacuating the gas from the assay bottles and replacing it with nitrogen (repeated at least 3 times) at 1 bar pressure and then maintaining the bottles at atmospheric pressure by releasing the pressure with a syringe. The required amount of concentrated solution of KNO_3 was added into all the anoxic assays for denitrification. The anaerobic assays were incubated at 37 ^{0}C on a shaker, whereas the anoxic assays were incubated at 25 $\pm$ 2 ^{0}C with shaking. All the assays were performed in duplicate.

2.3 Substrate medium

The synthetic wastewater used as the reactor influent in all the reactors was prepared in phosphate buffer of pH 7.3 (50 mmolar) and is described in Table 2.1. For the anaerobic fixed bed reactor treating 2-CP, the composition was slightly changed: 0.12 g/l yeast extract was used instead of meat extract and the amount of peptone and trace metal solution was reduced from 0.16 to 0.12 g/l and 1 ml/l to 0.2 ml/l, respectively. The components of wastewater were those of OECD synthetic wastewater with slight modification. Phenol and glucose was added in the reactors as mentioned, to maintain the initial influent COD of approximately 5 g/l for all reactors except for the anaerobic reactor treating 2-CP, where 2-CP was the only source of carbon (despite little peptone and yeast extract). The initial COD was 1 g/l. KNO_3 was added into the anoxic reactor to maintain a COD/NO_3 ratio in the feed from 3.4 – 4.

Table 2.1: Composition of the synthetic wastewater used as influent feed

Components	Amount
Peptone (g/l)	0.16
Meat extract (g/l)	0.11
Urea (g/l)	0.03
NaCl (g/l)	0.007
$CaCl_2·2 H_2O$ (g/l)	0.004
$MgSO_4·7 H_2O$ (g/l)	0.002
Trace metal solution* (ml/l)	1.0
Phosphate buffer (50 mM, pH 7.3) (ml)	1000

* Table 2.2

Table 2.2: Composition of the trace metal solution used for the influent feed (after Sarfaraz et al., 2004)

Components	Amount
$FeSO_4·7H_2O$ (g/l)	1.36
$Na_2MoO_4·2H_2O$ (g/l)	0.24
$CuSO_4·5H_2O$ (g/l)	0.25
$ZnSO_4·7H_2O$ (g/l)	0.58
$NiSO_4·6H_2O$ (g/l)	0.11
$MnSO_4·H_2O$ (g/l)	1.01
H_2SO_4 (ml/l)	1.0
Deionised water (ml)	1000

To study the effect of essential nutrients, the organic- (meat extract, peptone) and inorganic- (urea) sources of nitrogen and vitamins in the medium were replaced by only defined inorganic sources i.e. NH_4Cl 0.5 g/l and 1.33 ml/l vitamin solution (Table 2.3) as mentioned. This vitamin solution was stored in a dark bottle at 4 ^{0}C.

Table 2.3: Composition of the vitamin solution
(after Kafkewitz et al. 1996, modified from Wolfe's solution (Balch et al. 1979))

Components	Amount
Biotin (mg/l)	20
Folic acid (mg/l)	20
Pyridoxine hydrochloride (mg/l)	100
Riboflavin (mg/l)	50
Thiamine hydrochloride	50
Niacin (mg/l)	50
Pantothenic acid (mg/l)	50
Cyanocobalamin	10
p-Aminobenzoic acid	50
Thioacetic acid	50
Deionised water (ml)	1000

The medium used for the enrichment of aerobic cultures is described in Table 2.4

Table 2.4: Enrichment medium for aerobic cultures

Components	Amount
K_2HPO_4 (g/l)	0.5
$MgSO_4 \cdot 7\,H_2O$ (g/l)	0.2
$CaCl_2 \cdot 2\,H_2O$ (g/l)	0.01
NH_4NO_3 (g/l)	3.0
Trace metal solution* (ml/l)	1.0

* Table 2.2

Chapter 3

Results

3.1 Continuous reactor studies for phenol biodegradation

The degradation of phenol under continuous mode was studied in three suspended and two fixed bed reactors.

3.1.1 Suspended bed reactors

Three suspended bed reactors were started with the aim of acclimatising the microbial flora, obtained from municipal wastewater treatment plants to degradation of phenol under aerobic, anaerobic and anoxic conditions, respectively. The reactors were initially loaded with glucose as the sole carbon source at an initial COD of 5 g/l. Phenol was introduced in the reactor when the flora was well acclimatised to the high COD conditions.

3.1.1.1 Aerobic suspended bed reactor

The aerobic reactor was operated first with glucose as the sole carbon source for about 100 days at a COD of 5 g/l. The COD removal efficiencies of the reactor along with changes in HRT and OLR are presented in figure 3.1. The reactor was started with a HRT of 10 days and an OLR of 0.5 g/l·d. A week after the start, the COD removal reached > 84 %. No problems were encountered during glucose degradation in the reactor and it was able to degrade > 97 % COD at an OLR of 1.4 g COD/l·d resulting in a degradation rate of 1.36 g COD/l·d. The phenol was introduced in reactor influent on the 99th day of operation.

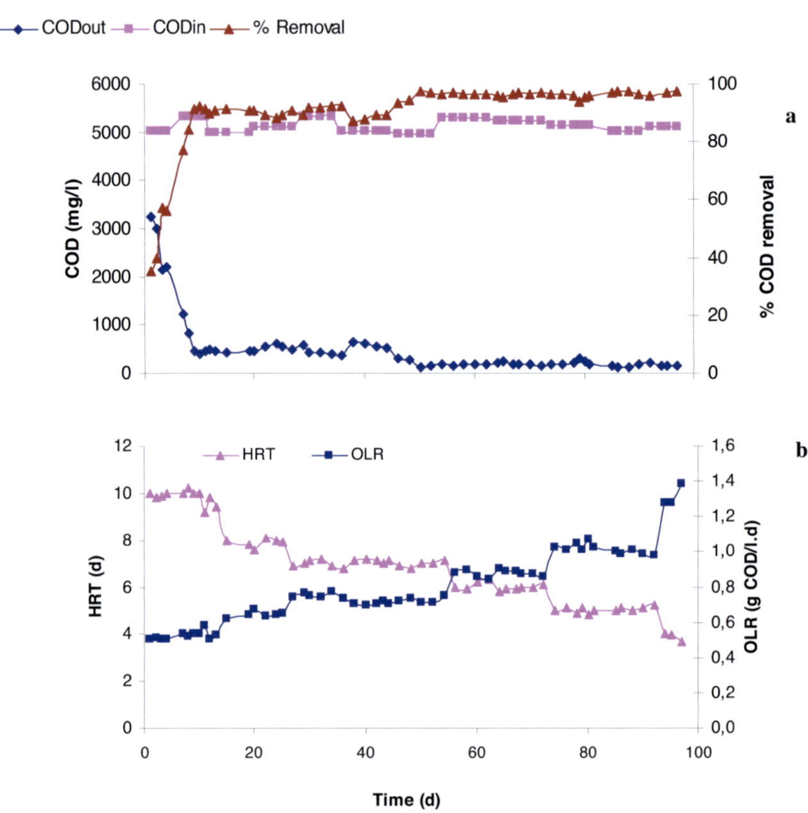

Fig. 3.1 – Performance of the aerobic reactor with glucose as the sole carbon source

a) COD removal and b) Changes in HRT and OLR

After introduction of phenol, the reactor operation was not very stable (Fig. 3.2). At a phenol concentration of 188 mg/l, the average removal efficiency was 93 %. However the COD removal was not parallel to the phenol removal as the former decreased from 97 % to 48 % within 30 days of phenol introduction in the reactor. The condition improved when phenol concentration was lowered from 188 mg/l to 94 mg/l.

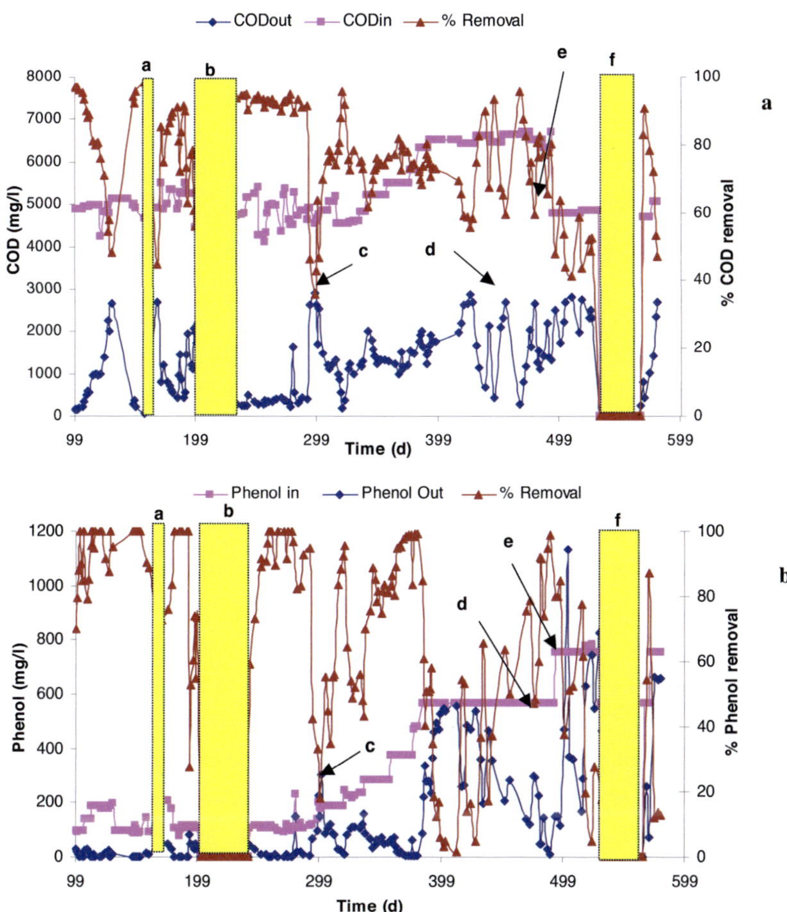

Fig. 3.2 – Performance of the aerobic reactor after introduction of phenol with the feed
a) COD removal and b) Phenol removal

Letter insert represent: a –oxygen deficient shock, b- no phenol period, c- increase in OLR, d- addition of new sludge, e- increase in phenol concentration (from 565 to 753 mg/l) and decrease in glucose (from 4 to 3 g/l), f- no feed period.

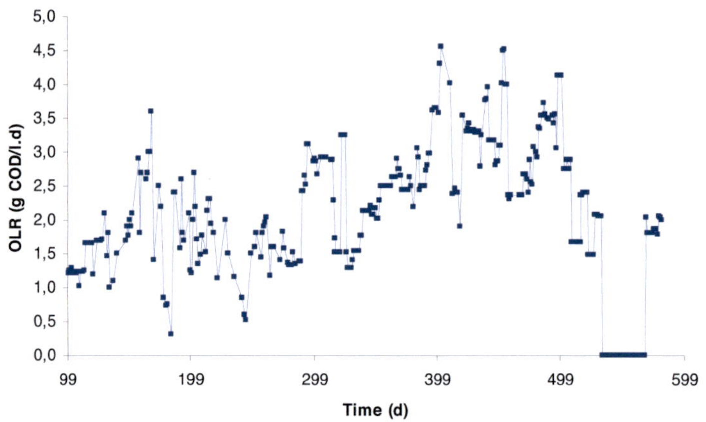

Fig. 3.3 – Organic loading rate after phenol addition in the influent

On day 164 of reactor operation (point a in Fig. 3.2), the air supply for the reactor was accidentally interrupted for 3 days. This caused a severe drop in COD removal (up to 44 %) and moderate drop in phenol removal (up to 72 %). More sludge was observed in the effluent and most of the sludge in the reactor changed to flocs. A portion of 200 ml fresh activated sludge from municipal wastewater treatment was added to the reactor and mixed thoroughly after the air supply was resumed. This action initiated better removal rates, but this condition was not stable for a long time. On day 193, COD and phenol removal showed a negative trail in efficiencies. COD was much affected as it came down to 56 %; similarly phenol removal was also unsteady. The colour of the reactor started to change from brown to yellow. Therefore on day 202 of reactor operation it was decided to stop phenol addition in the reactor influent (point b in Figure 3.2). The COD removal became quite stable and reached 95 %. Phenol addition at 104 mg/l concentration was resumed in the influent feed on day 242. On day 288 the OLR was increased from 1.4 up to 3.1 g COD/l·d till day 293 (Fig. 3.3). The reactor flora was unable to cope with the change in the loading conditions, which caused again a drop in the removal rates. This OLR of ca. 3 g COD/l·d was kept for 20 days and a maximum of 78 % COD removal and 56 % phenol (188 mg/l) removal was observed. In order to recover previous conditions, loading was decreased and > 95 % of both, COD and phenol removal were achievable within a week at an OLR of 1.52 g COD/l·d. As soon as the OLR was increased again to 3.2 g COD/l·d, a clear decrease in the efficiencies was observable with some sludge washout with the effluent. These changes in OLR demonstrated that the reactor could not bear fluctuations by almost a factor of 2 and the performance dropped down considerably at such conditions; however the performance could be improved when previous conditions were reversed.

From day 331 onwards the OLR was increased slowly instead of the abrupt changes. This approach proved to be suitable for better phenol degradation. The phenol concentration in the influent was also increased step wise. On day 383, the phenol concentration was increased up to 565 mg/l with 84.6 % removal at an OLR of 3 g COD/l·d. The reactor operation became very unstable and phenol removal dropped down to 1.7 % on day 411. The phenol removal improved to 54 % but it was again an unstable state. The colour of the reactor sludge turned from brown to pale yellow and a decrease in the COD removal efficiency was also observed. On day 479, fresh activated sludge (~ 400 ml) from the municipal sewage treatment plant, Karlsruhe was supplied to the reactor and it started to recover again. The phenol removal reached 95 % whereas COD removal was 77 % by day 486. On day 493, the concentration of phenol was further increased to 753 mg/l and glucose was decreased from 4 to 3 g/l (point e in Fig. 3.2). The changes in concentration once again lead to unstable reactor conditions. Even when the phenol concentration was lowered back from 753 to 565 mg/l, there was not a sign of improvement and phenol continued to accumulate in the reactor. Therefore no feed was supplied from day 533 to 566 (point f in Fig. 3.2). After restart of the reactor with 565 mg/l phenol, the removal efficiencies reached up to 87 % for phenol and 90 % for COD at an OLR of 1.8 g COD/l·d by day 570. When the concentration of phenol was increased to 753 mg/l, again the performance was affected and at the time of reactor termination on day 587, the phenol and COD removal was 9 % and 65 % at an OLR of 2.15 g COD/l·d and a HRT of 2.36 days.

3.1.1.2 Anaerobic suspended bed reactor

Parallel to the aerobic reactor an anaerobic suspended bed reactor was started with the aim of treating high COD containing wastewater, to which later on phenol was added. The reactor was fed for 140 days with glucose as the main carbon source in the feed. The reactor operation was divided into seven phases with respect to HRT and OLR during the time period when glucose was the only carbon source in the feed (Fig. 3.4).

The reactor was operated with the same substrate from a decreasing HRT from 10 days to 1 day and accordingly, an increase of the average OLR from 0.5 to 5 g COD/l·d was obtained. The maximum observed COD degradation rate was 5.3 g/l·d and the average removal rate from day 134 to 142 was 5.23 g COD/l·d. The mean COD removal during this period was 91.2 %. The increase of COD in effluent was related to the fatty acids accumulation in the reactor (Fig. 3.5).

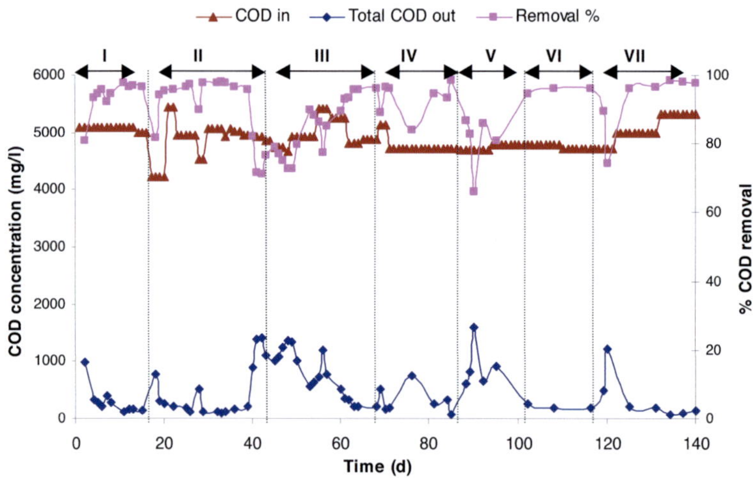

Fig. 3.4 – Performance of the anaerobic suspended bed reactor with glucose as the only source of carbon during different phases of the reactor operation.

The HRT (in days) was 10 (phase I), 8 (phase II), 5 (phase III), 4 (phase IV), 3 (phase V), 2 (phase VI), 1 (phase VII), corresponding to an OLR (in g COD/l·d) of 0.50 (phase I), 0.62 (phase II), 1.0 (phase III), 1.18 (phase IV), 1.58 (phase V), 2.37 (phase VI) and 5.07 (phase VII)

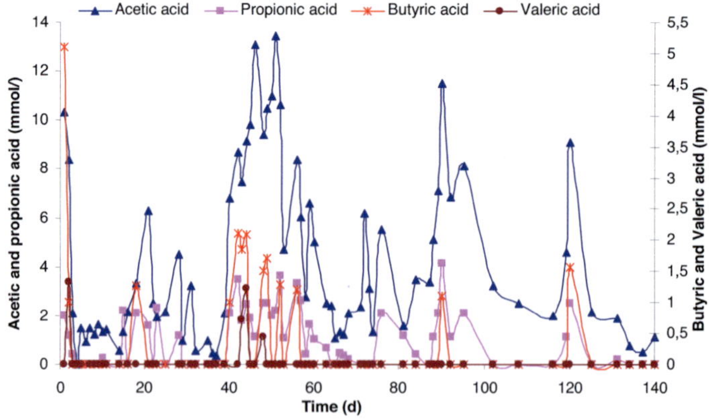

Fig. 3.5 – Fatty acid production with glucose as the only source of carbon in the reactor feed

The COD of each fatty acid was calculated and the sum of COD from the fatty acids corresponded for the total dissolved COD of the effluent (Fig 3.6).

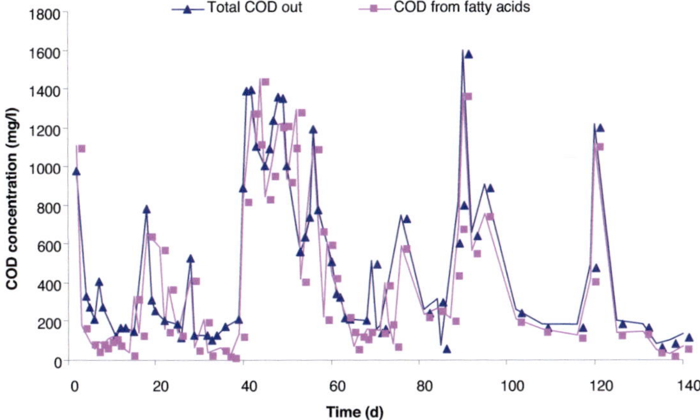

Fig. 3.6 – Total COD out and COD from fatty acids present in the effluent during degradation of glucose as the only source of carbon in the reactor feed

Biogas was readily produced in the reactor. However, the observed volume was always lower than the theoretical values (Fig 3.7), as calculated with the 'Buswell equation' (Equation 3.1) modified by Gallert and Winter (1999). The methane content of the biogas ranged from 40 % to 71.4 %

$$C_cH_hO_o + 1/4\ (4c\text{-}h\text{-}2o)\ H_2O \longrightarrow 1/8\ (4c\text{-}h\text{+}2o)\ CO_2 + 1/8\ (4c\text{+}h\text{-}2o)\ CH_4 \qquad \text{(Equation 3.1)}$$

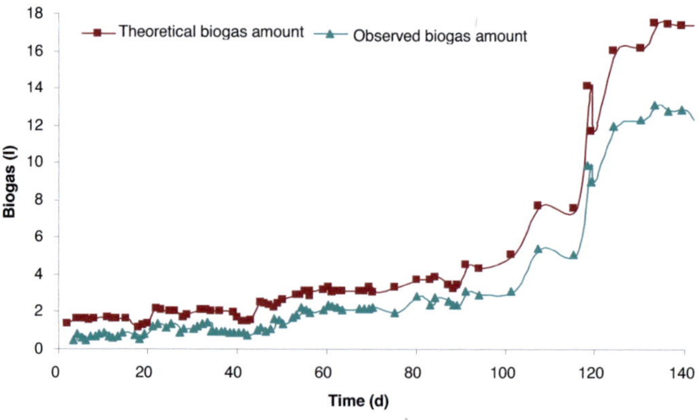

Fig. 3.7 – Biogas production with glucose as the only source of carbon in the influent

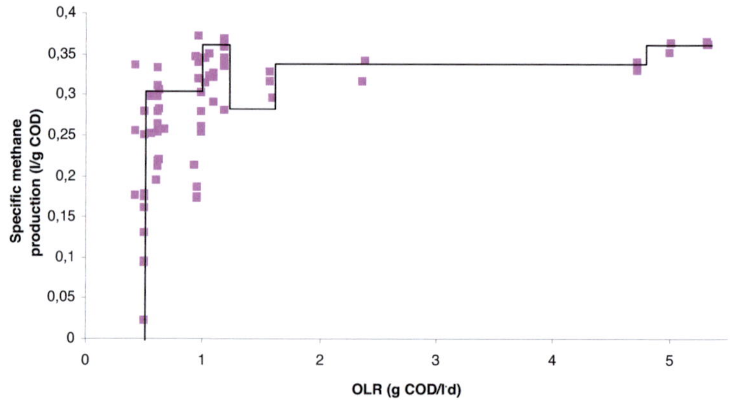

**Fig. 3.8 – Specific methane production with glucose as the
only source of carbon in the influent**

The expected specific methane production per unit of COD removal was 0.373. At the start-up the observed values were lower, but with time and with the gradual increase in OLR, methane production approached the theoretically expected values (Fig. 3.8).

On day 143 of reactor operation, phenol was added in the feed. The influent phenol concentration, changes in HRT and OLR during the entire operation are presented in Table 3.1.

Table 3.1: Loading pattern in the anaerobic suspended bed reactor during the entire operation period

Period	HRT (d)	OLR (g COD/l·d)	Phenol in (mg/l)
143-153	1.0	5.0	94
154-198	2.7	1.9	104
199-278	0	0	0
279-320	2.1	2.3	179
321-338	1.6	3.2	188
339-352	-	-	-
353-394	3.0	1.7	188
395-414	4.5	1.2	254
415-485	2.1	2.5	301
486-505	4.4	1.0	292
506-540	3.0	1.9	282
541-548	2.8	2.4	282
549-569	4.2	1.7	282
570-584	1.9	3.8	282
585-594	2.9	2.2	282
595-607	2.0	3.2	282
608-629	2.7	2.0	282.3
630-648	2.2	3.0	282 - 471
649-703	2.3	2.3	471 - 753
704-738	0	0	0
739-741	4.7	1.0	753
742-790	2.3	2.2	753 - 1129

The COD and phenol removal efficiencies are presented in Figure 3.9. At the time of phenol addition (94 mg/l) in the feed, the OLR was 5 g COD/l·d at a HRT of 1 day. Glucose was the major carbon source. The COD removal was around 90 % but the phenol removal was only 19 %. To improve the phenol removal efficiency, the HRT was increased from 1 to 3 days. Within 11 days, the phenol removal increased to 90 %.

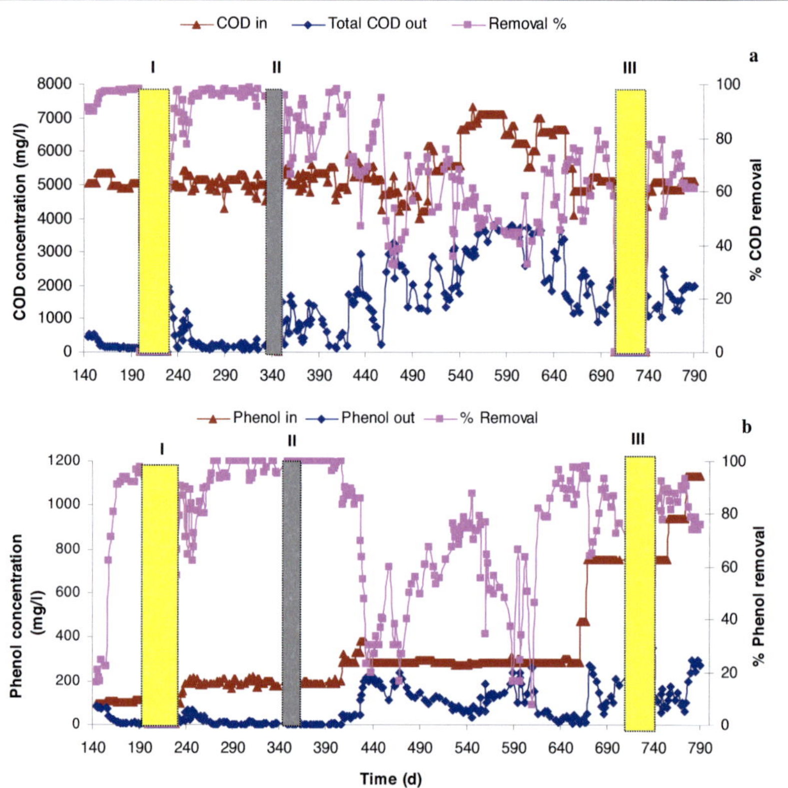

Fig. 3.9 – Reactor performance in terms of removal efficiencies after addition of phenol in the influent: COD removal a) and Phenol removal b)

Phase I: starvation period - no feed supply, phase II: batch mode and phase III: no feed supply

The reactor was operated for 1.5 months at an OLR of 1.9 g COD/l·d and a phenol concentration of 94 mg/l. On 198th day of reactor operation, 98 % COD and 97 % phenol were removed. In order to observe the effect of long term starvation, the reactor was not fed for one month from 198th day onwards. At the restart, it was capable to remove both, phenol and COD to about 60 %. In a week, removal efficiencies reached the levels as they had before the starvation period. The highest COD removal at an OLR of 5 g COD/l·d was 91 % with a phenol loading of 94 mg/l·d, of which only 24 % was degraded. The OLR, where 100 % phenol removal was observed was 3.4 g COD/l·d with a phenol loading of 132 mg/l·d and a HRT of 1.4 days on 322nd day of reactor operation. On this day, the observed biogas production was only 3.5 l whereas the theoretically calculated biogas production from the fermentation was 10 l (Fig. 3.10). The methane content of the biogas was 62 %, but still the total volume was less than the theoretically expected volumes.

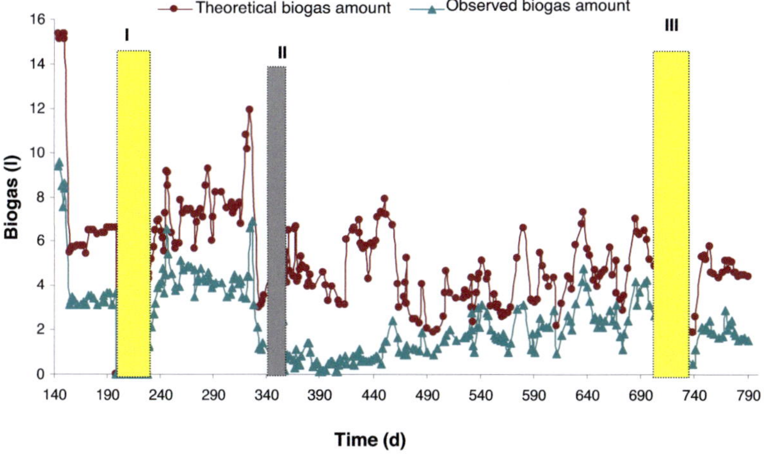

Fig. 3.10 – Theoretical and observed biogas in the anaerobic suspended bed reactor after phenol addition in the influent

Phase I: starvation period-no feed supply, phase II: batch mode and phase III: no feed supply

Some batch experiments to measure the phenol degradation rate were performed by taking out some sludge from the reactor. Whenever the sludge was taken out, it was replaced with fresh or stored sludge (not more than 2-3 weeks) of a municipal anaerobic reactor. On 330th day, 500 ml of the sludge were taken out and replaced mistakenly with biowaste sludge with a dissolved COD of about 60-70 g/l. The biowaste sludge had an adverse negative effect and it immensely decreased the specific methane production (Fig. 3.11).

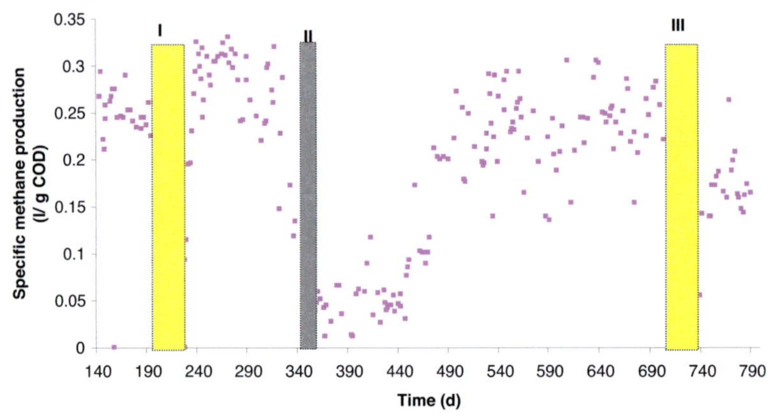

Fig. 3.11 – Specific methane production in the reactor

Phase I: starvation period-no feed supply, phase II: batch mode and phase III: no feed supply

Therefore from day 340-352, due to low biogas production, the reactor was operated in batch mode. During this mode, on full degradation of the substrate (a solution of glucose and phenol), consecutive feedings were provided to the reactor. The expected biogas production was low and was difficult to be measured with the gas meter. Therefore to estimate the exact biogas production, a water filled column (based on water displacement) was used to measure the gas volume. Both in presence and in absence of phenol the production of biogas observed was near to the theoretical values (Fig. 3.12). The substrate was provided 5-6 times, once the theoretical amount of biogas was experimentally observed. At all substrate feedings, the amount of biogas was always near to or exactly what was awaited. The reactor operation during batch mode ensured that the reactor sludge was fully active. After resuming the continuous operation, the COD removal was a little unsteady compared to the phenol removal.

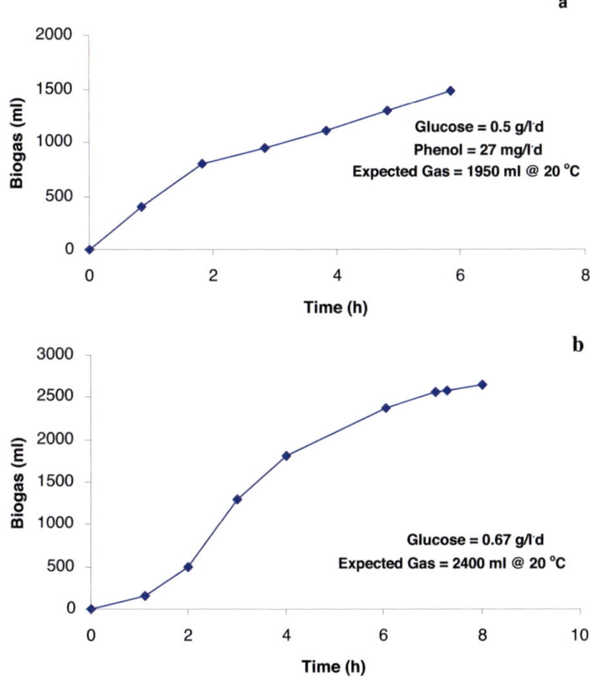

Fig. 3.12 – Biogas production during batch mode a) in presence of phenol
and b) in absence of phenol

On 422nd day of reactor operation, the warm water thermostat that was used for keeping the reactor at mesophilic conditions i.e. 37 ^{0}C, broke down. The reactor remained at temperatures between 15-20 ^{0}C for more than 40 hours. Due to this temperature shock, the COD removal dropped down from 95 % to 70 % and in case of phenol removal not an immediate but a gradual drop was observed. A corresponding decrease in fatty acids degradation was observed and fatty acids started to accumulate in the reactor (Fig. 3.13).

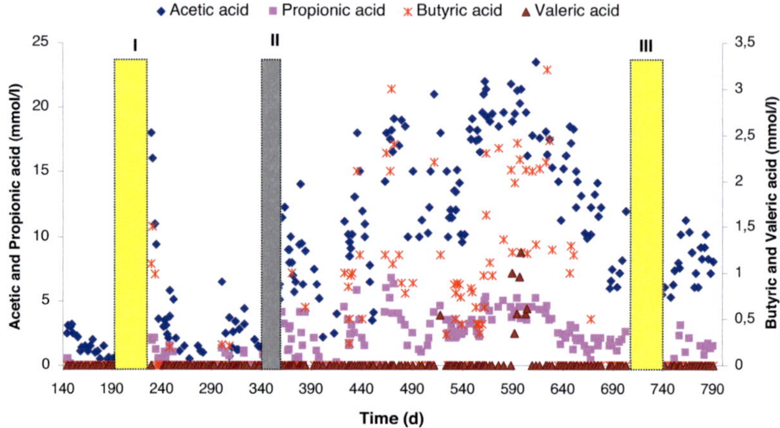

Fig. 3.13 – Fatty acid production in the reactor after phenol addition in influent

Phase I: starvation period-no feed supply, phase II: batch mode and phase III: no feed supply

The COD and phenol removal efficiencies continued to decrease. After a month COD removal was improved to 85 % but still the phenol removal was only 40 %. This condition was also temporary. The phenol removal improved to 82 % by the 616th day, but the COD removal was unsteady. The increase in phenol concentration from 282 to 471 and to 753 mg/l had no negative effect on the phenol removal efficiency but it effected COD removal, which remained to be low resulting in high concentration of fatty acids in the reactor liquor. The COD of each measured fatty acid and of the phenol was calculated as follows (Table 3.2)

Table 3.2: COD of phenol and fatty acids

Substance (1 mmol/l)	COD (mg/l)
Phenol*	224
Acetic acid	64
Propionic acid	118
Butyric acid	160
Valeric acid	208

*1 g/l phenol = 2.3 g/l COD

A comparison between the total dissolved COD and the COD from the sum of fatty acids and phenol together is presented in Figure 3.14.

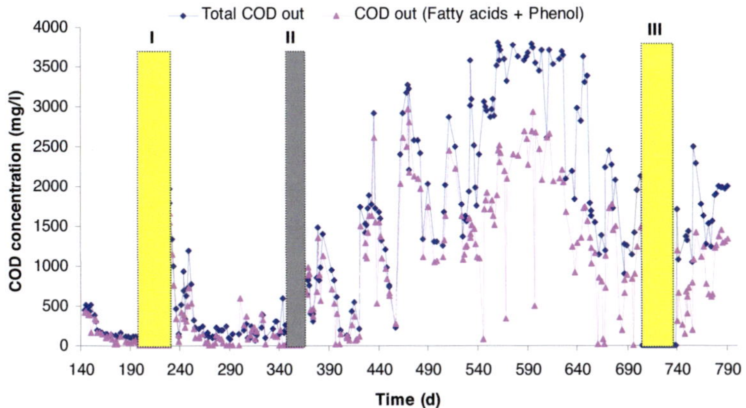

Fig. 3.14 – Total dissolved COD and calculated COD from fatty acids and phenol present in the effluent

Phase I: starvation period-no feed supply, phase II: batch mode and phase III: no feed supply

The total COD and the COD from residual fatty acids + phenol was quite comparable until day 532. The amount of COD measured was then more than the total amount of COD from the fatty acids and phenol in the effluent. On day 689, the COD removal started to decrease and reached to only 58 % by day 703. Therefore the reactor was not provided with any influent for one month. On day 739, when the reactor was restarted, 60 % of COD at an OLR of 1 g COD/l·d and 53 % of 753 mg/l phenol were degraded. The phenol concentration was gradually increased to 1129 mg/l. The highest phenol removal at this concentration was 90 % but it was not stable. At the time of reactor termination on day 790, the phenol removal was 76 % and the COD removal was 61 % at an OLR of 2.2 g COD/l·d and a HRT of 2.3 days. The highest COD removal rate observed after addition of phenol in the feed was 4.6 g COD /l·d on day 149, at an influent phenol concentration of 94 mg/l. Whereas, the highest phenol degradation rate observed was 437 mg/l·d on the 775th day of reactor operation at an influent phenol concentration of 1129 mg/l, an OLR of 2.18 g COD/l·d, a HRT of 2.34 days and an influent COD concentration of 5109 mg/l.

3.1.1.3 Anoxic suspended bed reactor

Parallel to the aerobic and anaerobic reactors, an anoxic suspended bed reactor was operated to investigate the continuous treatment of glucose and later on phenol as the co-substrate for long periods (766 days) under denitrifying conditions.

3.1.1.3.1 Enrichment of an anoxic culture

During the first 244 days of reactor operation, glucose was added as the sole carbon source into the influent wastewater to improve the proliferation of microorganisms. The entire operation for the enrichment of the sludge was divided into five phases of loading conditions (Table 3.3).

Table 3.3: Strategy for enrichment of an anoxic culture using glucose as the sole carbon source

Phase (d)	HRT (d)	OLR (g COD/l·d)	NO_3^--N (g/l)	C/N ratio
I (1-16)	10	0.5	1.56	3.2
II (17-29)	8	0.6	1.60	3.0
III (30-50)	5	1.0	1.60	3.1
IV (51-90)	3	1.7	1.50	3.4
V (90-244)	2.2	2.4	1.60	3.1

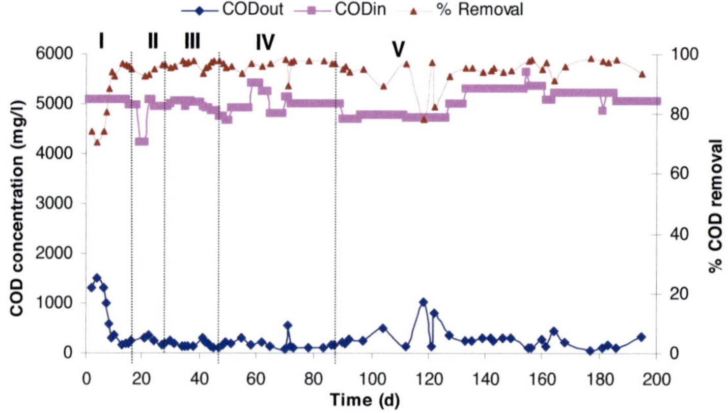

Fig. 3.15 – COD removal efficiency of the anoxic reactor during glucose degradation
See Table 3.3 for explanation of I, II, III, IV and V

The average total dissolved COD concentration during the first 200 days of the culture enrichment was 5018 mg/l. The removal ranged between 70 to 99 % with different HRT's and OLR's. The COD removal was 74 % on the first day of the reactor operation and within 2 weeks COD degradation increased to > 95 %. The highest COD removal rate was 2.5 g COD/l·d at an HRT of 2 days. The reactor was in steady state except for some short term fluctuations in removal efficiencies during phase V of the reactor operation (Fig. 3.15). From day 202-243 reactor effluent was not analysed, therefore it is not presented in figure.

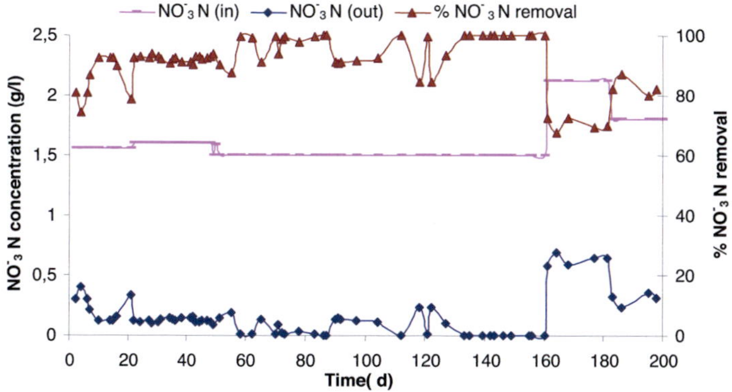

Fig. 3.16 – NO$_3^-$-N removal efficiency with glucose as the sole carbon source in the influent

Under anoxic conditions NO$_3^-$-N acts as an electron acceptor. Complete denitrification is the sequential reduction of nitrate (NO$_3^-$) to nitrogen (N$_2$) as follows:

$$NO_3^- \xrightarrow{2e^-} NO_2^- \xrightarrow{e^-} NO \xrightarrow{e^-} 0.5\,N_2O \xrightarrow{e^-} 0.5\,N_2 \qquad \text{(Equation 3.2)}$$

(Gallert and Winter, 1999)

Theoretically, reduction of 1 g of NO$_3^-$-N requires 2.83 to 2.86 g COD calculated for glucose or phenol as the substrates which can be deducted from the stoichiometry of equations 3.3, 3.4.

$$5\,C_6H_{12}O_6 + 24\,NO_3^- \longrightarrow 12\,N_2 + 24\,HCO_3^- + 6\,CO_2 + 18\,H_2O \qquad \text{(Equation 3.3)}$$

(Waste water technology, 1989)

$$5\ C_6H_5OH + 28\ NO_3^- + 28\ H^+ \longrightarrow 14\ N_2 + 30\ CO_2 + 29\ H_2O \qquad \text{(Equation 3.4)}$$

(Chakraborty and Veeramani, 2005)

The observed ratios during denitirification are however, higher as organic carbon is also utilised for the growth of biomass or cells. For the reduction of 1g NO_3^--N > 2.85 g COD is required (Bernet et al. 1996). Therefore, the influent COD/nitrogen (C/N) ratio was kept higher than 2.86. Along with the removal of COD, nitrate reduction was measured and is represented in Figure 3.16. The concentration of NO_3^-- N was kept constant except from day 160 to day 180 where it was increased to 2 g/l. This resulted in a decrease of nitrate reduction. However, the observed ratio of COD/nitrogen (C/N) removed remained constant (Fig. 3.17). The mean C/N removal ratio during the whole enrichment was 3.3.

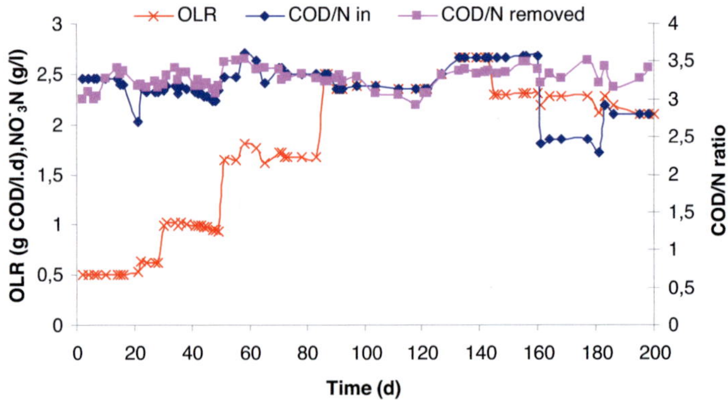

Fig. 3.17 – COD/Nitrogen (C/N) ratio at different organic loading rates

3.1.1.3.2 Reactor performance after phenol addition - COD and phenol removal

On the 246th day of reactor operation, 188 mg/l phenol was added to the influent along with glucose. The removal of phenol and COD was 72 and 88 %, respectively (Fig. 3.18). At an OLR of about 2.5 g COD/l·d, 100 % degradation of 188 mg/l phenol was observed during the 290th to 328th day of reactor operation. On day 328, the HRT was decreased to 1 day. This resulted in a steep decline in the phenol removal efficiency as only 45 % phenol degradation was observed at an OLR of 4.6 g COD/l·d. This condition was reversible as soon as the OLR was decreased by increasing the HRT to 2 days on the 339th day of reactor operation.

The phenol addition till 414th day was 207 mg/l and 98.6 % phenol was removed. As soon as the phenol concentration was increased from 207 mg/l to 329 mg/l, there was a drop in the COD removal efficiency from 94.8 % to 71.6 % and only 77 % phenol removal was observed. The COD removal efficiency improved gradually and it reached again 95.5 % on day 430. However the phenol removal efficiency remained around 75 %. When the phenol concentration was further increased, the COD removal decreased from 90.8 % to 81.1 % at 376 mg/l phenol and eventually reached 45.5 % with a corresponding decrease in phenol removal on day 450 for a phenol increment from 376 to 424 mg/l. When the phenol feed was lowered to 235 mg/l on day 451, the COD in the effluent decreased and the COD removal efficiency reached 96.7 % on the 468th day. This showed that the decrease of COD removal caused by increasing the phenol concentration was reversible.

On day 486, a drop in both, COD and phenol removal was observed for unknown reasons as neither OLR nor HRT or the phenol concentration was changed. However this was a short term fluctuation. By the day 489, the reactor was back at the previous conditions. But the reactor operation was not smooth. Frequent fluctuations in the removal efficiencies occurred. Most of the time, these fluctuations were partially or fully recoverable. The poor removal efficiencies were independent of OLR as decrease in OLR was not always helpful to stop the retrograding removal rates. The phenol concentration was increased gradually to 753 mg/l in the influent feed. It was observed that high removal efficiencies were achievable for both, COD and phenol (also at concentrations of 753 mg/l), but when the reactor was run longer at the given operational conditions, these efficiencies decreased. From day 710 to 745, no feed was added into the reactor to observe the changes after the long starvation periods. On day 745, when the reactor was started with 565 mg/l phenol in the feed at a HRT of 1.6 days and an OLR of 4.2 g COD/l·d, the removal was only 42 % and 10 % for COD and phenol, respectively. The reactor was operated further for 20 days and no remarkable differences in reactor performance were observable compared to the performance before the starvation period. The reactor operation was terminated on day 766 as the phenol started to accumulate. At the end of the operation, the removal of COD was 81 % with 207 mg/l·d phenol degradation rate at an OLR of 4.3 g/l·d COD and 1.1 days of HRT.

The maximum phenol loading in the reactor was 1412 mg phenol/l·d with a removal rate of 216 mg/l·d. Whereas the maximum phenol removal rate was 753 mg/l·d at a phenol loading of 828 mg/l·d observed on 694th day of the reactor operation with an OLR of ~6 g COD/l·d and a HRT of 0.9 days.

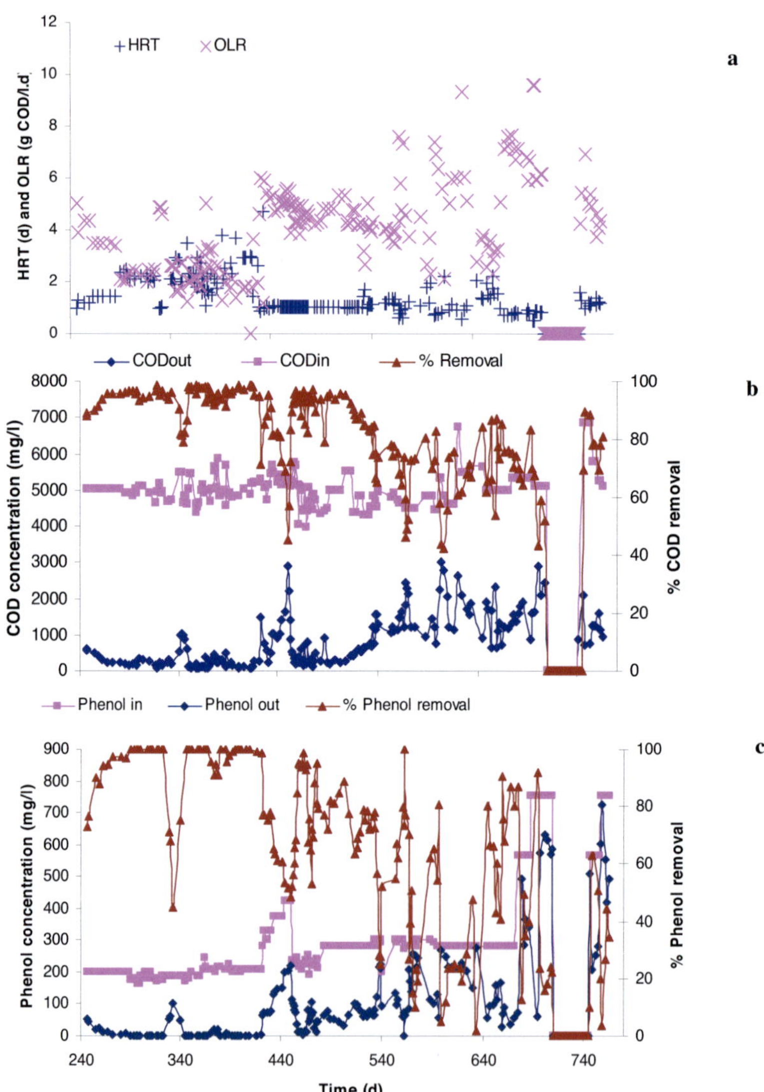

Fig. 3.18 – Reactor performance after phenol addition in influent, a) Different HRT and OLR during the reactor operation, b) COD removal, c) Phenol removal

3.1.1.3.3 Reactor performance after phenol addition- NO$_3$ reduction

The nitrate removal during the reactor operation is presented in Figure 3.19. The removal corresponded with the increase or decrease in COD.

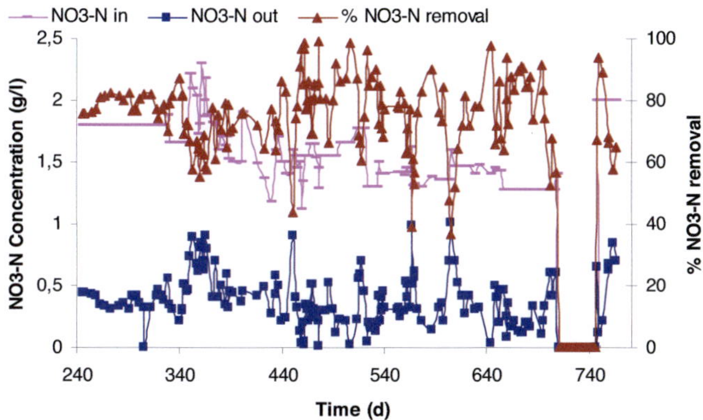

Fig. 3.19 – Nitrate-N removal during reactor operation after phenol addition in influent

The influent C/N ratio was kept higher as theoretically required also after phenol addition in the feed. The influent C/N ratio and the ratio of C/N removed are represented in Figure 3.20. From day 240 onwards, the observed mean ratio between removed COD and NO$_3^-$-N was 3.5.

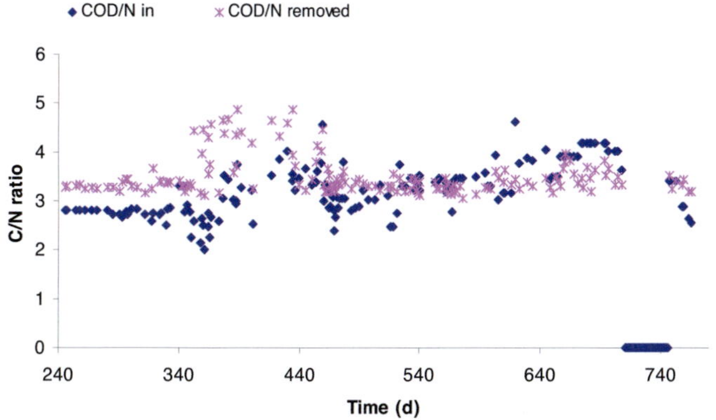

Fig. 3.20 – Carbon/ Nitrogen (C/N) ratios during the reactor operation

after phenol addition in influent

The net cell growth yield in the reactor was estimated by COD and nitrate removal expressed as nitrogen oxygen equivalent (NOE). NOE is expressed as the mass of oxygen that would accept the same number of electrons, when nitrate or nitrite is reduced to nitrogen gas. It is calculated using the following relationships:

$$\text{NOE consumption} = 2.86\ Q\ (N_1-N_2) + 1.71\ Q\ N_3 \qquad \text{(Equation 3.5)}$$

Q is the flowrate (l/d), N_1 is the influent NO_3^--N (mg/l), N_2 is the sum of the effluent NO_3^--N and NO_2^--N (mg/l) and N_3 is the effluent NO_2^--N (mg/l). The coefficients of 2.86 and 1.71 are equivalent to mg O_2 per mg NO_3^--N and NO_2^--N, respectively, which can be derived from the following reactions (Sarfaraz et al., 2004):

$$O_2 + 4\,H^+ + 4\,e^- \longrightarrow 2\,H_2O \qquad \text{(Equation 3.6)}$$
$$NO_3^- + 6\,H^+ + 5\,e^- \longrightarrow 0.5\,N_2 + 3\,H_2O \qquad \text{(Equation 3.7)}$$
$$NO_2^- + 4\,H^+ + 3\,e^- \longrightarrow 0.5\,N_2 + 2\,H_2O \qquad \text{(Equation 3.8)}$$

Equations 3.6 and 3.7 yield: $(32/4)/(14/5) = 2.86$ mg O_2/mg NO_3^--N

Equations 3.6 and 3.8 yield: $(32/4)/(14/3) = 1.71$ mg O_2/ NO_3^--N

This calculation is based upon the assumption that nitrate is the sole electron acceptor. The contribution from sulphate reduction is omitted as sulphate was present only in trace amounts in the influent (Tables 2.1 and 2.2) and the redox potential was too high for sulphate reduction. Typical light brown colour of the reactor sludge and the absence of the characteristic odour produced from sulphate reduction were taken as an indication that no sulphate reduction had occurred. Gas production was measured and occasionally methane was found in the reactor, but only in traces (1-2 %). No or minimal nitrite (10-15 mg/l) was observed in the effluent. Therefore contribution of NO_2^--N to NOE is not considered for the calculation and only electrons consumed for nitrate reduction are considered for the NOE. As previously described, the average C/N ratio observed for the reactor operation after addition of phenol in the feed was 3.5. NOE can be calculated by dividing 2.86 by 3.5, which is 82 %, the residual 18 % COD contributed to sludge growth as well as to the suspended solids wasted in the effluent. The daily NOE observed after phenol addition is presented in Figure 3.21.

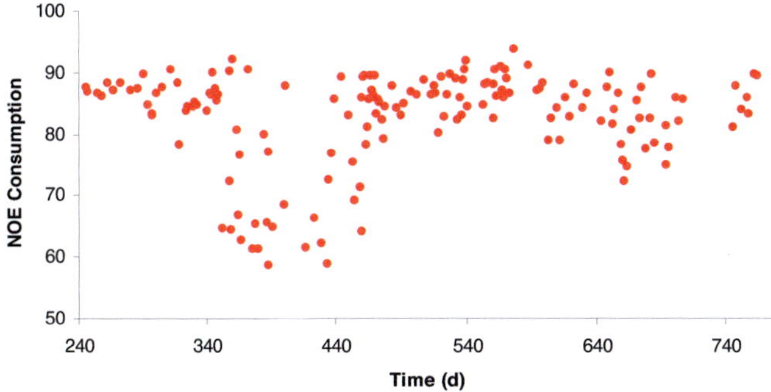

Fig. 3.21 – Nitrogen oxygen equivalents (NOE) during the reactor operation after phenol addition

3.1.2 Fixed bed reactors for phenol biodegradation

In order to achieve degradation of high strength phenolic wastewaters under aerobic and anaerobic conditions, studies were carried out in fixed bed reactors.

3.1.2.1 Aerobic fixed bed reactor

The effluent of the suspended bed reactor (Sec. 3.1.1.1) was used as an inoculum for the start-up of an aerobic fixed bed reactor with a feed containing both, glucose and phenol. The concentration of phenol was increased continuously. The operation period of this reactor was 403 days.

3.1.2.1.1 Reactor performance in terms of phenol removal

The OLR and phenol loading followed during the entire reactor operation are mentioned in Table 3.4.

Table 3.4: Phases during the reactor operation

Days	Avg. OLR (g/COD l·d)	Avg. dissolved influent COD (g/l)	Phenol Conc. (g/l)
1-60	2.2	5.0	0.19 - 0.28
61-120	4.9	6.0	0.28 –0.94
121-180	5.4	8.2	0.94 – 2.54
181-240	5.4	12.6	2.63 – 4.70
241-291	5.0	14.8	4.70 –5.17
292-310	7.1	14.9	4.89
311-323	13.7	14.9	4.89
324-354	9.7	14.9	4.89
355-405	7.3	14.9	4.89

Reactor performance at the varying conditions is shown in Figure 3.22 (a-f). The initial HRT was 3.6 days at an OLR of 1.67 g COD/l·d. The phenol concentration in the influent was 0.19 g/l and only 17 % of phenol was removed on the first day. Within the first week, there was a visible increase in sludge growth on the surface of the carrier material and 96 % of phenol removal was achieved. Therefore the HRT was decreased and the OLR was increased further. The phenol concentration was further increased to 3.95 g/l in 0.094 to 0.188 g/l increments within 7 months of reactor operation (Fig 3.22 b). Before increasing the phenol to 4.7 g/l in an increment of 0.753 g/l, the OLR was ca. 6 g COD/l·d with 99 % phenol removal. The removal efficiency dropped to 89 % at an influent phenol concentration of 4.7 g/l. This condition was improved to 94 % phenol removal within a day, when the OLR was decreased by 0.5 g COD/l·d. The phenol supply was further increased up to 5.17 g/l and the reactor was operated for one month. The phenol removal gradually decreased to 93 %. To stop a further decline, the concentration of phenol was decreased to 4.89 g/l. This showed an increase in removal up to 98 %. The OLR was then increased slowly to 8.6 g COD/l·d. It resulted in only 59 % phenol removal. When the OLR was increased further, the removal efficiency decreased.

The highest phenol removal rate in the reactor was observed on Day 313 of reactor operation. It was 2.93 g phenol/l·d at a HRT of 0.95 day and a total OLR of 15.3 g COD/l·d with 4.89 g/l of phenol loading in the feed. However the reactor performance was not stable. For improvement, the loading was decreased to 12 g COD/l·d but still the phenol removal was < 50 % at this loading. The OLR was then decreased further to ca. 7.5 g COD/l·d and the reactor was operated for 3 days at this loading. The OLR was again increased to 10.4 g COD/l·d. The phenol removal started to improve. However phenol removal at this OLR was only about 72 %. To get better removal rates, the OLR was again decreased. Phenol removal improved continuously and reached 96 % at an OLR of 7.5 g COD/l·d with a phenol removal rate of 2.3 g/l·d on day 403 of reactor operation.

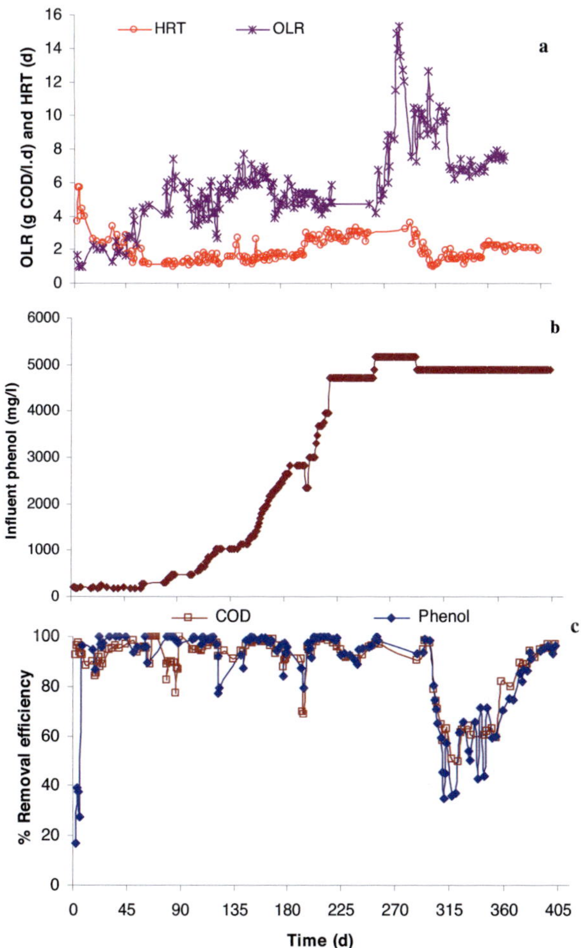

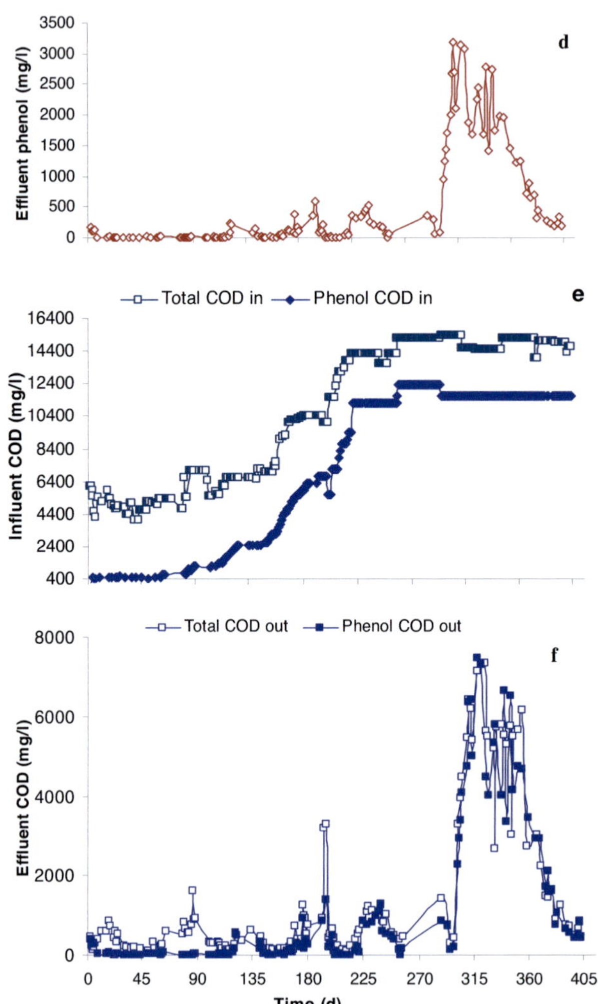

Fig. 3.22 – Performance of an aerobic fixed bed reactor: a) HRT and OLR, b) Influent phenol concentration, c) Phenol and COD removal efficiencies, d) Effluent phenol concentration, e) Influent COD and f) Effluent COD concentration. On day 219, glucose was reduced from 5 to 3 g/l in the feed and the reduced COD supplemented with phenol (1 g Phenol = 2.38 g COD).

3.1.2.1.2 Reactor performance in terms of COD removal

According to the oxidation stoichiometries, the COD of 1 g glucose is 1.06 g (Equation 3.9) and the COD of 1 g phenol is 2.38 g (Equation 3.10).

$$C_6H_{12}O_6 + 6\,O_2 \longrightarrow 6\,CO_2 + 6\,H_2O \qquad \text{(Equation 3.9)}$$

$$C_6H_5OH + 7\,O_2 \longrightarrow 6\,CO_2 + 3\,H_2O \qquad \text{(Equation 3.10)}$$

At the start of the reactor, the main source of COD in the influent was glucose. In the initial stages of reactor operation, the contribution of COD from glucose and other nutrients was 5 g/l and only ~ 470 mg/l came from phenol (Fig. 3.22 e). The COD concentration was increased along with each phenol increment. By the day 170 of reactor operation, phenol was the abundant carbon source and glucose was successively replaced as the main substrate. As can be seen from Figures 3.22 c and f, the sharp decrease in COD removal always corresponded with increasing phenol concentration in the effluent. This indicated that COD in the effluent was due to residual phenol rather than glucose.

On day 219, glucose was decreased from 5 g/l to 3 g/l and phenol addition was increased to keep the COD constant. The COD removal slowly decreased during the next 25 days and phenol in the effluent increased. After a short recovery phenol degradation fell to less than 40 % due to an increase of the OLR to about 15 g COD/l·d. It successively recovered to 98 % COD removal within 90 days at 4.89 g/l phenol in the influent and an OLR of 6 - 7 g COD/l·d after doubling the HRT. The ratio of phenol to total $COD_{dissolved}$ at that stage was 0.75. At the end of the reactor operation on day 405, the total dissolved influent COD was 14.7 g/l, of which 11.6 g/l was contributed by 4.89 g/l of phenol (Table 3.5). The effluent phenol concentration was 0.188 g/l, contributing to 0.44 g/l dissolved COD, which exactly corresponded to the total effluent COD observed (0.44 g/l). It is obvious that the effluent COD was due to residual phenol and not due to other non-degraded carbon sources, such as glucose, peptone or meat extract in the medium. The ratio of phenol COD to total dissolved COD was 0.78 at the termination of the reactor.

Table 3.5: Balance of phenol and COD concentrations in the fixed bed reactor at day 405

Parameter	Values
Total COD in (g/l)	14.7
Total COD out (g/l)	0.44
Phenol COD in (g/l)	11.6
Phenol COD out (g/l)	0.44
Phenol COD/Total dissolved COD	0.78

3.1.2.1.3 Effect of peptone and meat extract on phenol degradation

The reactor was provided with a feed solution containing glucose (5 g/l until day 218, then 3 g/l), peptone (0.16 g/l), meat extract (0.11 g/l) and urea (0.03 g/l) along with inorganic nutrients and increasing amounts of phenol (Fig. 3.22 b). On day 218 after start of the reactor > 97 % of 13.8 g/l total dissolved COD, including 9.4 g phenol-COD/l, was aerobically degraded. The phenol removal was 99 %, equivalent to 3.95 g/l·d, the OLR at this stage was 5.7 g COD /l·d. On day 219, the concentration of phenol in the influent was increased to 4.7 g/l, glucose was decreased from 5 to 3 g/l and no peptone, meat extract and urea were added in the feed. This resulted in an increase of the effluent phenol and COD concentrations. On day 239, the phenol effluent concentration reached 0.51 g/l (Fig. 3.23) with 90.8 % removal of 14.2 g COD/l (COD_{phenol}: $COD_{total\ dissolved}$ = 0.79). At this point peptone, meat extract and urea addition to the feed were resumed. This caused a significant improvement in phenol and COD removal within the next 15 days. The average HRT during this study period was 2.8 days and the average organic loading was 5 g COD/l·d, respectively. On day 255, removal of 97 % COD (14.2 g/l) and 99.8 % of phenol (4.7 g/l) were observed.

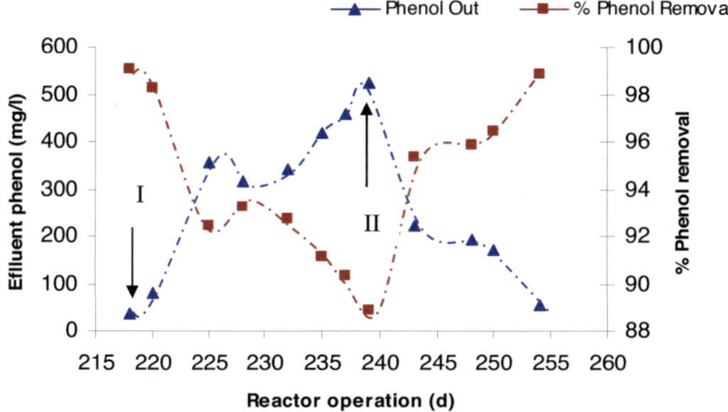

Fig. 3.23 - Effect of peptone, meat extract and urea on phenol removal

I-Omission of peptone (0.18 g/l), meat extract (0.11 g/l) and urea (0.03 g/l) in feed

II-Resumption of peptone, meat extract and urea addition in feed

3.1.2.2 Anaerobic fixed bed reactor for phenol degradation

The effluent of the anaerobic suspended bed reactor (Section 3.1.1.2) was used as an inoculum for start-up of the anaerobic fixed bed reactor with a feed containing both glucose and phenol. The concentration of phenol was increased continuously and at stable conditions a phenol shock was applied. Operating conditions and removal efficiencies of the reactor are described in Table 3.6. The reactor was operated for 550 days, and details are presented in the following sub-sections.

Table 3.6: Reactor performance before, during and after phenol shock load

Days	Avg. OLR (g COD/l·d)	Avg. total COD $_{in}$ (mg/l)	Theoretical phenol COD $_{in}$ (mg/l)	Avg. total COD removal (%)	Phenol $_{in}$ (g/l)	Avg. phenol removal (%)	Remarks
0-6	0.86	4909.5	448	68.3	0.19	31.2	
7-21	1.0	4745.4	448	58.5	0.19	83.0	
22-38	1.2	4651.9	448	72.0	0.19	67.0	
39-63	0.8	4250.7	672	92.7	0.28	89.2	Steady state
64-73	1.28	5325.0	672	97.7	0.28	85.0	reactor
74-81	-	0	0	-	0	-	performance at
82-84	1.43	5345.9	672	88.8	0.28	94.7	continuous
85-88	1.70	6474.7	896	93.6	0.37	99.8	increase in phenol
89-92	1.8	6065.1	1120	88.7	0.47	99.2	loading
93-95	-	0	0	-	0	-	(Phenol and
96-121	1.8	6357.5	2016	87.5	0.85	97.8	glucose (4 g/l) served as carbon
122-159	2.6	6641.5	3136	96.3	1.31	97.2	sources in feed)
160-184	3.6	9537.2	6272	99	2.63	99.4	
185-204	4.2	11570.0	7840	98.5	3.3	98.8	
205-221	5.3	13237.9	8960	89.2	3.76	93.7	
222-223	5.7	14199.0	11200	91.6	4.70	89.5	
224-231	2.6	14199.0	11200	74.4	4.70	70.4	
232	1.65	14199.0	11200	0	4.70	41.0	Decrease of
233-237	1.42	14199.0	11200	-	4.70	43.1	glucose to 3g/l
238-244	-	0	0	-	0	-	and increase of
245-246	0.55	14150.0	11200	11.6	4.70	21.0	phenol by 0.94 g/l resulted in reactor
247-248	-	0	0	-	0	-	performance
249-250	3	14150.0	11200	19.0	4.70	18.0	disturbance
251-261	2.5	14150.0	11200	26.2	4.70	24.3	
262-294	-	0	0	-	0	-	
295-298	0.55	4707.9	1344	0	0.56	0	
299-303	-	0	-	-	-	-	
303-305	-	0	-	-	-	-	
306-321	1.65	5006.0	1792	52.6	0.75	27.0	Efforts done to restore the former
322-324	1.54	4000.0	0	42.5	0	-	reactor
325-326	-	0	0	46	0	-	performance by
327-333	1.34	3700.0	672	48.3	0.28	26.8	adding phenol in
334-342	1.61	4103.0	0	53.3	0	-	the feed at small
343-345	2.72	4287.0	0		-	-	concentrations
346-347	1.7	4287.0	0		-	-	preceded by no
348-357	1.37	4180.5	0	60.3	-	-	phenol in the
358-361	1.23	4314.7	448	74.6	0.19	78.0	influent.
362-365	1.66	5460.0	1344	80.8	0.56	45.9	
366-372	2.20	5837.1	1792	94.2	0.75	83.6	
373-378	2.08	6281.0	2240	81	0.94	65.6	
379-386	1.85	6638.7	2688	87.2	1.13	79.2	
387-417	2.55	6758.0	2688	80.4	1.13	71.6	
418-443	3.29	6717.0	2688	73.17	1.13	78.7	
444-450	1.96	6781	3136	85.1	1.31	88.2	
451-481	1.71	3334.6	3136	60.2	1.31	90.0	
482-513	1.34	3286.0	3136	76.42	1.31	90.6	Phenol as the only
514-523	3.0	3286.0	3136	76.34	1.31	80.0	carbon source in
524-550	3.5	3108.0	2688-3136	47.5	1.13 – 1.31	52.6	the feed medium

3.1.2.2.1 Microbial acclimatisation to high phenol concentrations in the anaerobic reactor

The reactor was started at an organic loading rate (OLR) of 0.85 g COD/l·d. The phenol concentration in the influent was only 188 mg/l (Fig. 3.24 a). Only 21 % phenol was removed on the first day after start up with pre-adapted sludge. The removal reached > 94 % within one week of operation. On the 13th day, the gas head was changed, which caused air intrusion in the reactor; therefore the phenol removal rates dropped from 92 % to 56 %. Also, the OLR was slightly increased to about 1.2 g COD/l·d. Recovery was observed within 2 weeks, and phenol removal reached around 80 %. To improve the efficiency, the OLR was again decreased to 0.85 g COD/l·d and an average of 90 % phenol removal was achieved thereafter. Then the phenol load was increased to 283 mg/l, accompanied by an OLR increase to 1.3 g COD/l·d. The phenol degradation efficiency was on average 85 %. To demonstrate the stability of the process, on day 74 of reactor operation, no feed was provided for 1 week. When feeding of the reactor was resumed again, > 90 % phenol removal was still achievable. Also, interruption of feeding for 3 days did not have any negative effect on phenol removal (e.g. day 93-95). At a phenol concentration of 376 mg/l, the OLR was increased to 1.7 g COD/l·d. The phenol removal was constant at > 98 %. The phenol concentration was further increased to 3.76 g/l in small increments within 7 months of reactor operation while operating the reactor at 3.76 g/l influent phenol concentration for 16 days, the performance was stable and on average 94 % phenol removal was observed at such a high phenol concentration. During this period, the maximum phenol removal was 3.7 g /l at a hydraulic retention time (HRT) of 2.5 days and an OLR of 5.3 g/l·d, phenol loading of 3.5 g/l·d which amounted to a phenol removal rate of 1.45 g /l·d.

3.1.2.2.2 Phenol shock loading and recovery

On day 221 of reactor operation, the phenol concentration was increased from 3.76 g/l to 4.7 g/l, the glucose concentration was decreased from 4.2 to 3 g/l and no peptone, meat extract and urea were provided in the influent. These changes caused a decrease in phenol removal (Fig. 3.24 b) from 98.5 % at 3.76 g phenol/l to 89 % at 4.7 g phenol/l in one day. In order to prevent a further decline of the reactor performance, the OLR was decreased from 5.7 to 2.64 g COD/l·d by increasing the HRT from 2.5 to 5 days, but leaving the phenol concentration in the influent at 4.7 g/l. This did not result in efficiency improvement. Within the next two weeks phenol removal dropped to 43 % in spite of further lowering the OLR to 1.4 g COD/l·d. Phenol accumulation in the reactor began on day 237. No feed was then provided to the reactor for 8 days. On day 245 feed was resumed with 4.7 g phenol/l at an OLR of 0.5 g COD/l·d and peptone, urea and meat extract were added in the feed as before the shock loading. The phenol removal was, however, only 21 %. So again, no feed was provided for the next two days. Feed supply was started again on day 249, the OLR was kept at around 2.5 g COD/l·d and the reactor performance was observed for the next 13 days. The average phenol removal at 4.7 g/l was 24.3 %. To improve the conditions, the reactor was not provided with any feed for approximately one month. After one month, on day 295 influent was supplied again to the reactor with only 0.56 g/l phenol at 0.5 g COD/l·d, but there was still a problem of phenol accumulation. The influent medium was then replaced with only tap water as feed for 4 days in order to dilute the accumulated phenol. Then 200 ml supernatant from an anaerobic reactor treating municipal sewage, stored at 37 ^{0}C for three months, was provided to the reactor. Tap water was withdrawn from the reactor and the microbial flora was allowed to set up for 3 days in the reactor without any feed. After 3 days, influent with 0.75 g/l phenol was given into the reactor at an OLR of 1.7 g COD/l·d. The average phenol removal for the next 5 days was approximately 39 %. From day 311, phenol degradation efficiency continuously decreased for an unknown reason and on day 321 phenol accumulation in the reactor was observed. On day 322 the reactor was reintroduced with 500 ml of supernatant from an anaerobic reactor at the municipal sewage plant, Karlsruhe. For the next 4 days no phenol was added with the reactor feed. Phenol at 0.28 g/l concentration was added into the influent on day 327 until day 333, but only 27 % phenol was removed. In order to stabilise the biomass and to grow the sludge in the reactor, no phenol was added in to the influent for next 24 days. On day 358, influent was provided with 0.19 mg/l phenol and the reactor showed 78 % phenol removal within 4 days. When the removal increased, the concentration in the medium was also increased in small increments, up to 1.12 g/l by day 379. At each concentration increment, there was a temporary decrease in removal which improved as soon as the microbial flora was acclimatised to the higher concentration. The reactor was operated with 1.12 g/l phenol for 2 months. On day 443, 96 % phenol removal was observed at an OLR of 2 g COD /l·d and a HRT of 3 days. The peptone, meat extract and urea in the influent were then replaced by 0.5 g/l NH$_4$Cl and a vitamin solution. On day 443, phenol in the medium was increased to 1.31 g/l. After 7 days of operation, the removal rate increased up to 99 %. On day 450 of reactor operation, no glucose was provided in the feed and phenol was the main carbon source for the biomass. The average phenol removal was 90 % at an average OLR of 1.5 g COD /l·d until day 512 of reactor operation. On day 513, the HRT was decreased from 2.2 to 1.1. This resulted in a decrease in phenol removal and during a further decrease of the HRT to 0.9 days, the observed average phenol removal was only 54 %. So, the phenol concentration in influent was decreased from 1.31 to 1.12 g/l. At the end of the investigation, on day 550 phenol removal was 58 % of 1.12 g /l phenol.

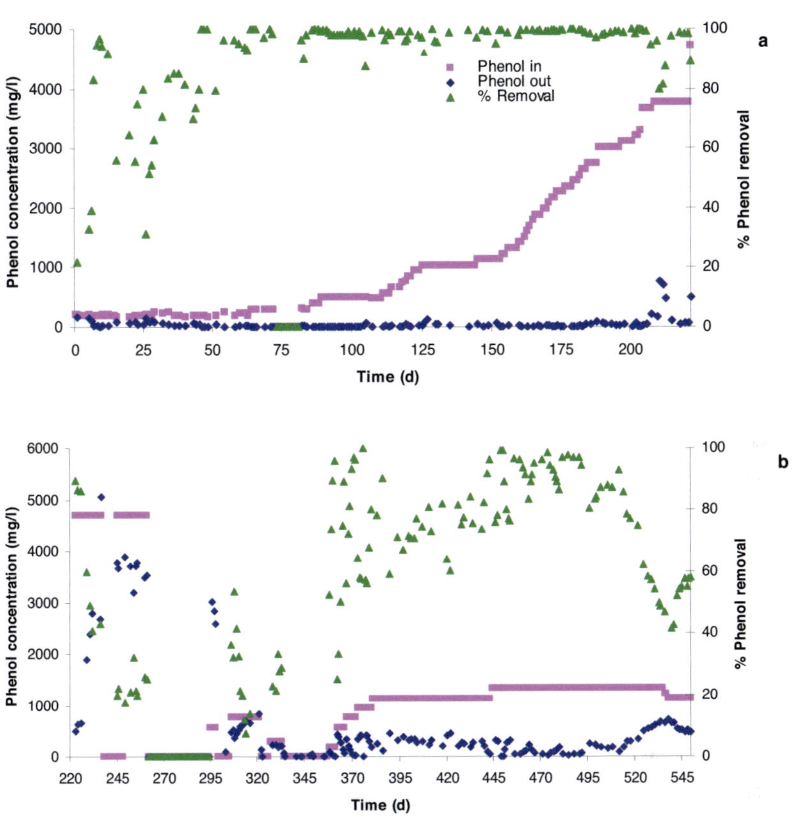

Fig. 3.24 - Phenol removal efficiency of the reactor from day 0-222 (a) and from day 222 onwards (b)

3.1.2.2.3 COD removal before and after phenol shock loading in an anaerobic fixed bed reactor

The reactor was started with approximately 5 g/l COD (Fig. 3.25 a). The COD from phenol contributed only 450 mg/l at the start of the reactor. Most of the COD came from glucose (4 g/l) and other nutrients. The average COD removal was 66 % for a month after reactor start up. This removal rate was improved continuously and reached 97.4 % on day 48 of reactor operation. There was only a short term decrease in COD removal efficiency when the changes in operating parameters (i.e. either HRT, OLR or small increases in phenol loading) were made; the COD efficiency in general was always recoverable and satisfactory (Fig. 3.25 a). From day 74 onward, no feed was supplied to the reactor for one week. After resuming feeding, the removal rate was still 89 %. The COD removal corresponded to the phenol removal indicating that no phenol intermediate accumulated in the reactor and phenol was being mineralised completely. The COD removal at 3.76

g/l phenol concentration i.e. before exposing the reactor to the shock load was 93.4 %. at an OLR of 5.3 g COD/l'd.

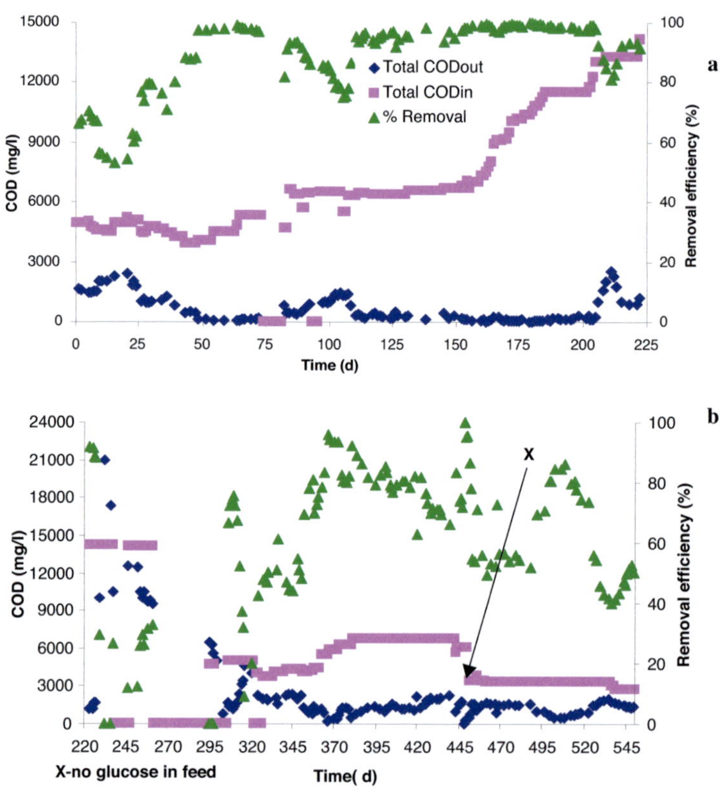

**Fig. 3.25 - COD removal efficiency of the reactor from day 0-222 (a)
and from day 222 onwards (b)**

On day 221, the changes in glucose and phenol concentration in the feed caused a steep decline in COD removal efficiency and after 8 days, only 29.5 % COD removal was observed (Fig. 3.25 b). Accumulation of non-degraded phenol/glucose in the reactor totally prevented COD removal. To allow for recovery, no feed was provided for one week on day 237. After that only ≤ 32.5 % COD was removed until day 265. Therefore feed was stopped for 1 month. The reactor was fed after one month with influent containing glucose and only 0.56 mg/l phenol as carbon sources. There was still COD accumulation, so feed was discontinued and the reactor was provided with municipal sewage effluent. The feed was started again with 0.75 mg/l phenol; the COD removal was 66 % but then decreased gradually to 8 % on day 316. For further improvement in COD removal, no phenol was added to the influent and glucose was again increased from 3 to 4 g/l on day 334. Glucose remained the main carbon source in the feed for 23 days. The removal rate reached 80 %. Then, phenol was again introduced into the reactor at a concentration of 0.19 mg/l on day 358 and some decline in COD removal was observed which quickly recovered. The phenol concentration was increased gradually to 1.31 g/l in the next 2.8 months. On day 450, the glucose supply in the influent was stopped and phenol became the only carbon source in the feed. This had a negative effect on COD removal efficiency. The phenol removal was greater than the COD removal. The mass distribution (Fig. 3.26) between the total dissolved COD and fatty acids + phenol in the effluent indicated that there was an accumulation of intermediate products of anaerobic phenol metabolism, e.g. benzoate, if degradation proceeded via the 4-hydroxybenzoate pathway. But this was temporary and by day 498, COD removal almost matched the phenol removal, indicating that the removed phenol was being completely degraded until the end of the experiment.

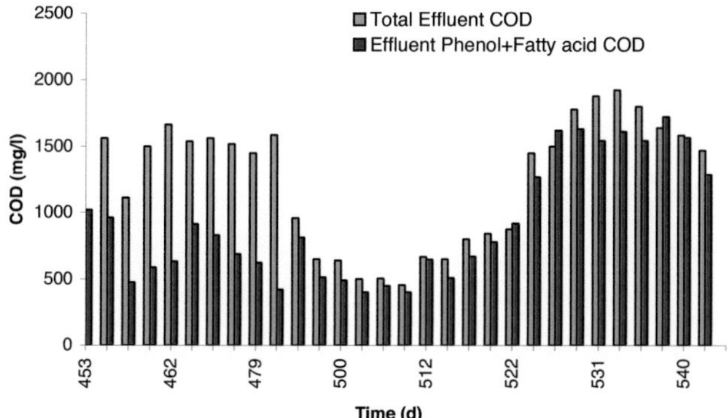

Fig. 3.26 - Effluent total and phenol/fatty acid-imparted COD concentrations during operation with phenol as the sole carbon source

3.1.2.2.4 Fatty acids and biogas production

The production of fatty acids before and after the reactor start-up is presented in (Fig. 3.27 a). A high fatty acid production, especially of acetic acid, was observed in the first 22 days of reactor start-up, after which fatty acids began to decline. Almost no fatty acid production was then observed until day 204. At day 204, the phenol concentration was increased from 3.30 to 3.67 g/l. This caused some build-up of fatty acids, which lasted up to day 220. After phenol shock loading from day 223 onwards, fatty acids started to accumulate in the reactor (Fig. 3.27 b). By the day 233, acetic acid reached 25 mmol/l. When no feed was provided to the reactor, the concentration of fatty acids decreased. As soon as feeding was started (on day 292) after a period of one month starvation, fatty acids were present in the effluent. The lowering of the phenol concentration did not contribute to fatty acid reduction, as 20 mmol/l acetic acid in the effluent was measured with 0.75 g/l (8 mmol/l) phenol and 3 g/l glucose in the feed at an OLR of 1.6 g COD/l·d. Avoidance of acid accumulation was only achieved when feeding was stopped for several days. On day 334, no phenol in the feed was provided and glucose was increased back to 4 g/l from 3 g/l. Immediately after this action, fatty acid accumulation was observed. When phenol was provided as the sole carbon source on day 450, a moderate build up of acetic acid was observed for some time. This concentration of fatty acids was not sufficient to account for the difference between total dissolved effluent COD and the COD from the effluent phenol (Fig. 3.26). At the end of reactor operation no fatty acids were found in the effluent.

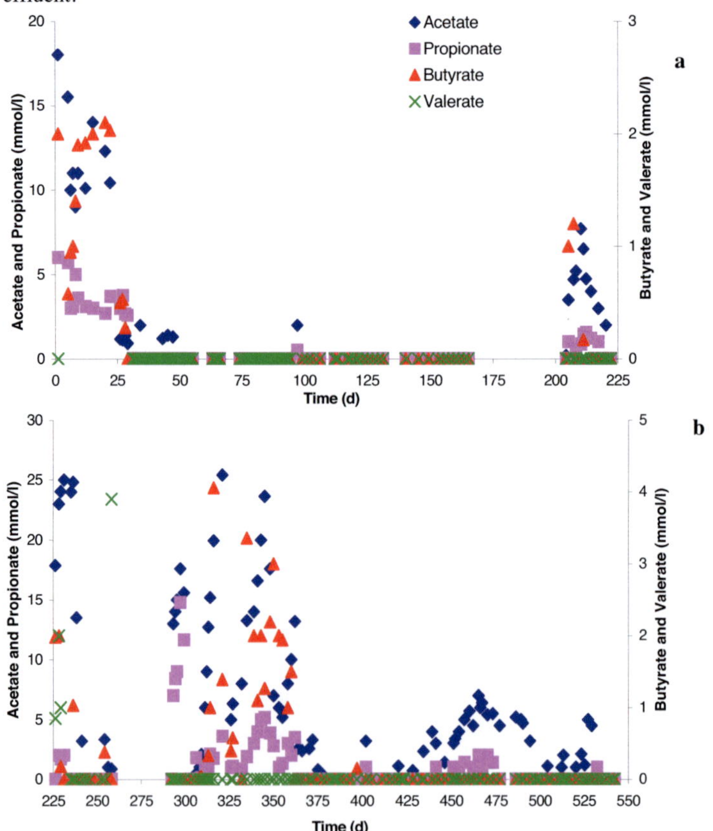

Fig. 3.27 - Fatty acids in the reactor effluent from day 0-222 (a) and from day 222 onwards (b)

The production of biogas is represented in Figure 3.28. The observed biogas quantity was less than the expected amount as calculated from equation 3.1 (Section 3.1.1.2). The quality of gas before shock loading was very good as the methane content was always 60-75 % of the total gas. No hydrogen and gases other than CH_4 and CO_2 were observed during gas chromatographic analyses of the biogas. During shock loading, the production of biogas was very low and after shock loading the quality of gas in terms of methane production decreased. The methane content of the biogas after shock loading and during recovery was between 40-60 % due to CO_2 release from the bicarbonate buffer system at decreasing or lower than normal pH.

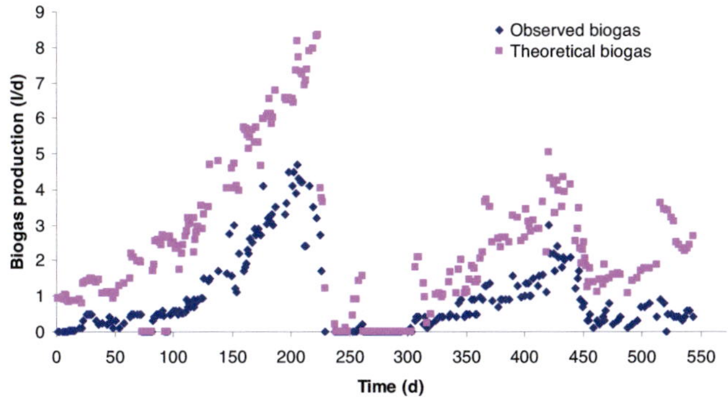

Fig. 3.28 - Biogas production throughout the reactor operation

The average specific methane production (SMP) observed before shock loading was 0.20 l CH_4/g COD removed, whereas the maximum SMP observed was 0.36 l CH_4/g COD removed on day 176 at an OLR of 3.8 g COD/l·d and an influent phenol concentration of 2.26 g/l. This value of SMP was very close to the theoretical value of 0.37 l CH_4/g COD. The average value of SMP after shock load was 0.15 CH_4/g COD. The maximum SMP observed after shock load was 0.26 CH_4/g COD on day 428 at an OLR of 3.8 g COD/l·d and an influent phenol concentration of 1.13 g/l. The methane production throughout the reactor operation is presented in Figure 3.28.

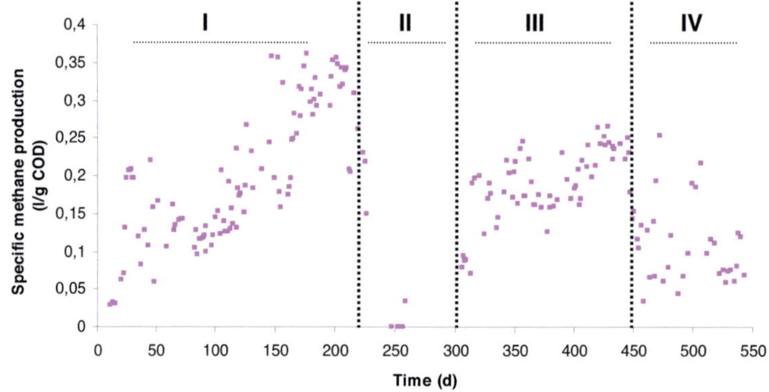

Fig. 3.29 – Specific methane production throughout the reactor operation

Phase I: increasing loading, phase II: Overload, phase III: recovery period and phase IV: phenol as the sole carbon source.

3.2 Batch studies of phenol biodegradation

A number of batch experiments were carried out with effluent of the continuous reactors to determine the degradation behaviour under the respective reactor conditions. Some of the important results are presented in the following sections.

3.2.1 Aerobic batch assays

Aerobic batch assays were carried out to study the phenol degradation profile and the effect of co-substrates on phenol acclimatised and unacclimatised sludges. The biokinetic growth parameters were also determined under aerobic conditions.

3.2.1.1 Phenol degradation profile

The phenol degradation profile was determined in a batch experiment by taking sludge from the aerobic fixed bed reactor on day 232 to study the trend of degradation rates over a short period of time and the effect of addition of an easily degradable co-substrate such as acetate to the same batch. The VSS concentration of the 10 % sludge after dilution in the enrichment culture was 3.5 g/l. A phenol removal of 31.5 % was observable within 11 h of incubation, corresponding to 4.7 mmol/l or 0.45 g/l phenol. After 30 h of incubation 93.4 % of 15 mmol/l (1.4 g/l) phenol was degraded (Fig. 3.29).

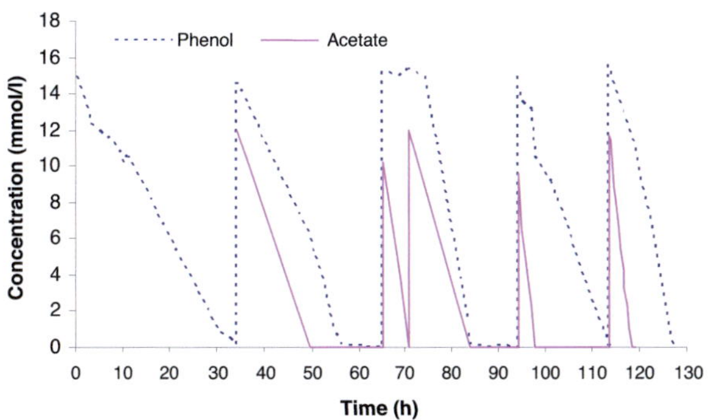

Fig. 3.30 – Phenol degradation profile during consecutive feedings in a batch assay

1 mole phenol = 94.1 g

To the same batch, 1 g/l of sodium acetate was added along with 15 mmol/l (1.4 g/l) phenol. The culture had a pH of 6.8 and was kept overnight at 25 ^{0}C on a mechanical shaker (120 rpm). After 15.5 h, 8.5 mmol/l (0.8 g/l) phenol was degraded. After 24 h of incubation, 99.5 % phenol was removed which accounted for 14.6 mmol/l (1.37 g/l) phenol. It showed that addition of a co-substrate had a positive effect on phenol biodegradation. The pH of the above batch increased from 6.8 to 7.7. This is because acetate was consumed leaving behind the cation sodium as sodium hydroxide. To the same batch, again 15 mmol/l (1.4 g/l) phenol and 10 mmol/l acetate was added and the pH was adjusted to 6.8. The volume was adjusted to 50 ml by addition of mineral medium. Complete acetate removal was observed within 5.5 h after its addition. At this point another 1 g/l sodium acetate was added to the batch. After 19 h, the pH of the medium was 8.1 and 99 % phenol was degraded. Addition of more co-substrate (double amount of the previous run) had no negative effect on phenol removal. It was also observed that 15 mmol/l (1.4 g/l) of phenol was removed in a shorter period. Thus the bacterial activity increased due to acclimatisation. This batch was once more fed with 15 mmol/l phenol and 10.5 mmol/l acetate. The degradation time was again 19.5 h for phenol and 4 h for acetate. In the following re-feeding cycle of 15 mmol/l of phenol and 11.2 mmol/l of acetate, it took only 13 h for 98 % phenol removal (Fig. 3.30 a). The time requirement for degradation of a certain amount of the phenol continued to decrease, mainly due to a shorter lag-phase (Fig 3.29). The highest phenol degradation rate observed was 1.7 mmol/l·h equal to 160 mg/l·h or 3.84 g/l·d. The highest acetate removal rate was 5.38 mmol/l·h equal to 322 mg/l·h or 7.74 g/l·d (Fig. 3.30 b). The TS concentration at the end of the experiment was 9.27 g/l and the VSS concentration was 5.32 g/l.

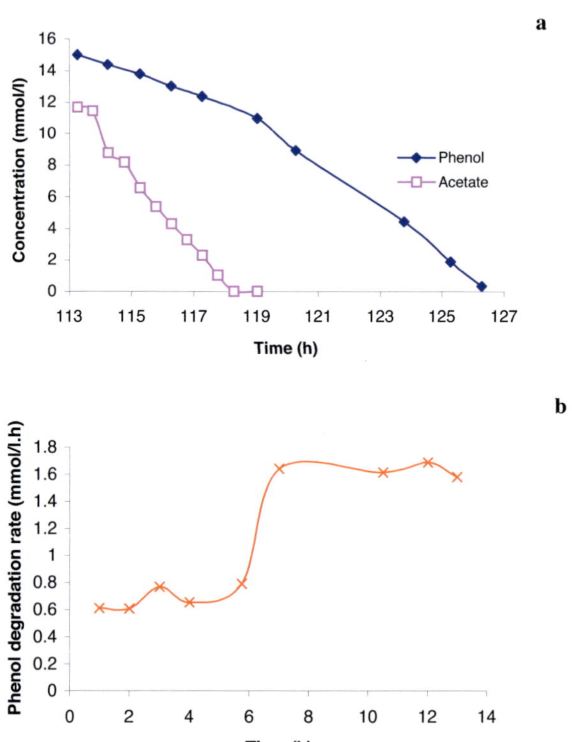

Fig. 3.31 - Phenol degradation profile at the fifth consecutive feeding (a) and phenol degradation rate (b)

3.2.1.2 Batch analysis – Substrate co-metabolism with acclimatised sludge

Degradation tests of phenol $\pm$ co-substrates such as glucose, acetate and 2-chlorophenol (2-CP) were performed with sewage sludge that was well adapted to phenol degradation and cultured in batch mode.

3.2.1.2.1 Degradation of phenol as the sole carbon source

This set of experiment was performed with a phenol-acclimatised microbial consortium subjected to suitable proliferating conditions by repeated inoculation into enrichment medium with phenol as an external carbon source (10 mmol/l or 0.94 g/l). Enrichment medium with 7 mmol/l (0.65 g/l), phenol as the sole carbon source was inoculated with 10 % of the above phenol acclimatised sludge.

Phenol degradation was analysed and upon depletion, 7 mmol/l were replenished once again (Fig. 3.31 a). The removal of the first 7 mmol/l phenol took about 5 h at a degradation rate of 42.5 mmol/l·d (4g/l·d). The degradation of the second amount of 7 mmol/l phenol proceeded at a somewhat better rate of 51.6 mmol/l·d (4.85 g/l·d).

3.2.1.2.2 Effect of acetate on phenol removal

Profiles of the degradation of acetate and phenol by phenol acclimatised sludge in enrichment medium, containing 7 mmol/l phenol and 7 mmol/l acetate are shown in Figure 3.31 b. The data indicate that phenol biodegradation was retarded in the presence of acetate as compared to assays with a phenol as a sole carbon source (Fig. 3.31 a). It therefore appears that the presence of acetate has a significant effect on the phenol utilization activity by sludge acclimatised to phenol as the sole carbon source with consecutive feedings over a long time. It took approximately 7 h for the complete phenol removal, whereas acetate was utilised completely within 4.25 h. The acetate utilisation rate was faster than that of phenol. At the beginning a lag phase in the phenol utilisation is evident when acetate was present. Acetate is one of the intermediate of phenol degradation and may cause a feed-back inhibition. Phenol degradation began when acetate was depleted to less than 2 mmol/l.

3.2.1.2.3 Effect of glucose on phenol removal

Phenol acclimatised sludge showed a preference for glucose respiration and utilised phenol only after glucose was depleted (Fig. 3.31 c). The presence of glucose attenuated the rate of phenol removal more than in the case of acetic acid as a second substrate (Fig. 3.31 b). As soon as glucose was respired, a rapid increase in phenol degradation was observed at a rate of 41.6 mmol/l·d (3.9 g/l·d) and 92.3 % phenol was removed within 6.25 h, most of it in the last 2 hours of incubation (Fig. 3.31 c).

3.2.1.2.4 Effect of 2-CP addition

The effect of 2-CP on phenol degradation is illustrated in Figure 3.31 d. Due to the high toxicity of 2-CP, only 1.2 mmol/l (0.15 g/l) of 2-CP was added along with 7 mmol/l phenol to the batch assay. Phenol biodegradation was significantly retarded by 2-CP. Including an initial lag-phase of almost 12 h, it took more than one day for the complete removal of phenol. When phenol was degraded, 2-CP was slowly utilised during the following 130 h. A second addition of 1.1 mmol/l 2-CP did not lead to an improved 2-CP degradation rate (Fig. 3.31 d).

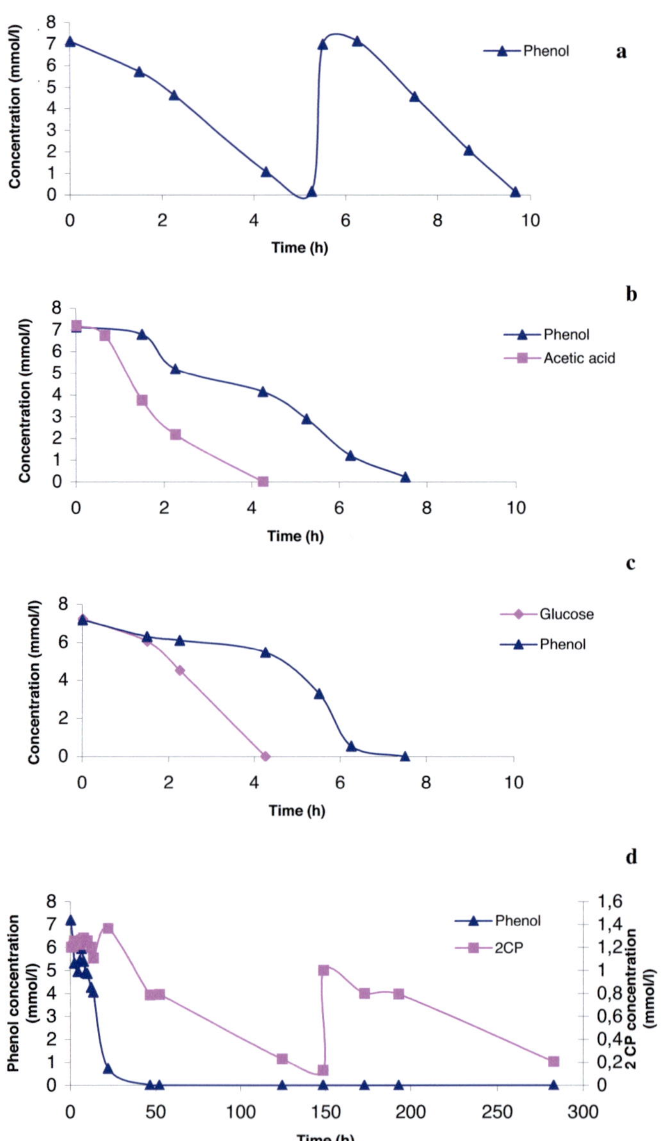

Fig. 3.32 - Phenol degradation by acclimatised sewage sludge (a) and phenol degradation in presence of acetate (b) glucose (c) and 2-chlorophenol (d)

3.2.1.3 Batch analysis – Substrate co-metabolism with unacclimatised sludge

This set of experiments was carried out with fresh aerobic activated sludge taken from the municipal wastewater treatment plant, Karlsruhe to study phenol degradation ± co-substrates.

3.2.1.3.1 Phenol Degradation

Four sets of batch assays were carried out to determine the effect of phenol on non-acclimatised sewage sludge. Phenol (2 mmol/l or 0.19 g/l) was added in the fresh sludge taken from a municipal wastewater treatment plant for the assays I-IV (Table 3.7). Assays II, III and IV were additionally supplemented with acetate (4 mmol/l, assay II), glucose (4 mmol/l, assay III) and 2-CP (1 mmol/l, assay IV), respectively. There was no phenol removal in any of the assays during first 8 h (data not shown). After 24 h, no phenol was detected in assay I, II, and III, whereas neither phenol nor 2-CP were removed in the assay IV (Table 3.7).

The utilisation of 2 mmol/l phenol in non-adapted sludge within 24 h would be quite unusual as phenol is not an easily degradable organic substance. However, the high SAC_{254} and COD values of the samples from all assays after 24 h indicated that phenol was not mineralised to CO_2 and H_2O, but was presumably converted to an aromatic intermediate (Table 3.7), which was not detectable by gas chromatography during phenol measurement and no phenol or other peaks were observed after 24 h. It could have been benzoate.

Table 3.7 : Phenol and co-substrate degradation using non-acclimatised sludge

Assay	Sample	Phenol T=0 h (mmol/l)	Phenol T=24 h (mmol/l)	SAC_{254} T = 0 h	SAC_{254} T=24 h	COD T=0 h (mg/l)	COD T=24 h (mg/l)	% COD removal
I	Phenol	2.0	0	1.760	1.293	485.52	362.47	25.3
II	Phenol + A[a]	2.0	0	1.772	1.390	740.75	383.8	48.2
III	Phenol + G[b]	2.0	0	1.796	1.416	1214.24	412.24	66.0
IV	Phenol + 2-CP[c]	2.0	1.98	2.470	2.421	697.9	685.4	1.8

Where: a) A = Acetate, b) G = Glucose, c) 2-CP = 2-Chlorophenol

3.2.1.3.2 Co-substrate degradation

In assay II, acetate was completely degraded after 6 h, similarly as seen in sludge adapted to phenol, but at a slower rate (data not shown). The degradation of glucose as a co-substrate together with phenol (assay III) was very slow. No degradation was observed during the first 9 h, but glucose was completely degraded after 24 h. In the case of phenol and 2-CP feeding (assay IV), no removal of either phenol or 2-CP was observed, even after 10 days of incubation (data not shown).

3.2.1.4 Enrichment culture from adapted sludge

A mixed bacterial culture was enriched from the phenol adapted sludge by 7 consecutive transfers (each 5 %) in enrichment medium with phenol as the sole carbon source. This selective enrichment culture served as an inoculum for the assays described in Figure 3.32. All assays contained 6 mmol/l (0.56 g/l) phenol from the beginning and were conducted simultaneously under identical conditions. To each of the duplicate assays, 6 mmol/l glucose or acetate, respectively, were added 6 h after the start of incubation at the onset of phenol degradation (Fig. 3.32, arrow) and phenol removal as well as glucose and acetate concentrations were determined (Figs. 3.32 and 3.33). It is interesting to note that phenol-acclimatized enrichment cultures had a preference for phenol and did not utilize glucose. It is also evident that the presence of glucose did not influence phenol degradation, which proceeded at the same rate as in the assays without a co-substrate (Fig. 3.32).

The presence of acetate, on the other hand, significantly slowed down the rate of phenol degradation. Phenol-adapted cells apparently preferred acetate over phenol. The suspended cultures grew very homogenously in all three assays and the OD values increased from $A_{578} = 0.14$ to 1.1. Biomass growth corresponded with phenol and/or acetate utilization. The externally added acetate (6 mmol/l), which was completely degraded, served as a preferential growth substrate and reduced phenol utilization (Fig. 3.33). If one assumes 3 mmol/l of acetate as an intermediate of phenol respiration, then 1.5 mmol/l of non-degraded phenol (at 11.5 h, Fig. 3.33) would be equivalent to 4.5 mmol/l acetate. These were not required since 6 mmol/l acetate were available from the addition of acetate.

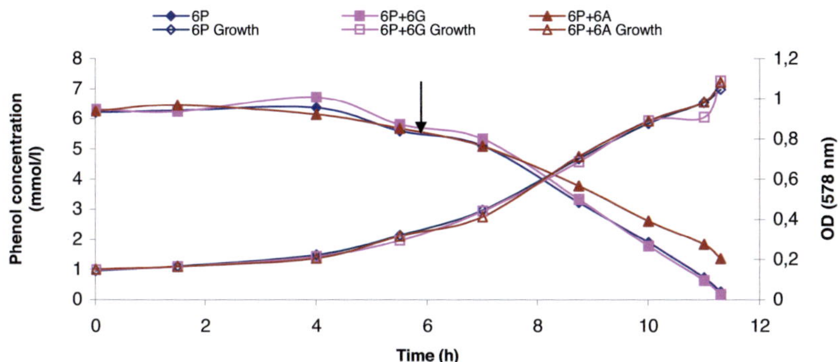

**Fig. 3.33 – Phenol degradation and growth of biomass in enrichment culture
from phenol adapted sludge**

Assay 6P contains phenol as sole carbon source, whereas in the assays 6P + 6G and 6P + 6A, 6
mmol/l glucose and acetate, respectively, were added 6 h after the start of the incubation (see
arrow).

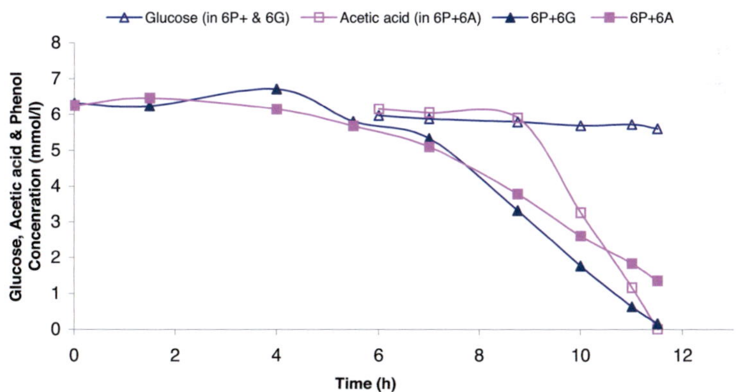

**Fig. 3.34 - Phenol and co-substrates degradation in enrichment cultures
from phenol adapted sludge**

1 mmol phenol = 94.1 mg

3.2.1.5 Kinetics of phenol biodegradation

Cell growth and phenol degradation at concentrations ranging from 0.25 to 7 mmol/l (23.5 to 658.8 mg/l) were measured as a function of time (Fig. 3.34). The lag phase of growth was longer with increasing initial phenol concentrations (Fig. 3.34 a). The degradation of phenol as a function of time is shown in Figure 3.34 b. The initial rate of phenol degradation for 3, 4 and 7 mmol/l (282.3, 376.4 and 685.8 mg/l) phenol concentration in the assays was linear for 3, 4 and 6 h, respectively, indicating substrate inhibition.

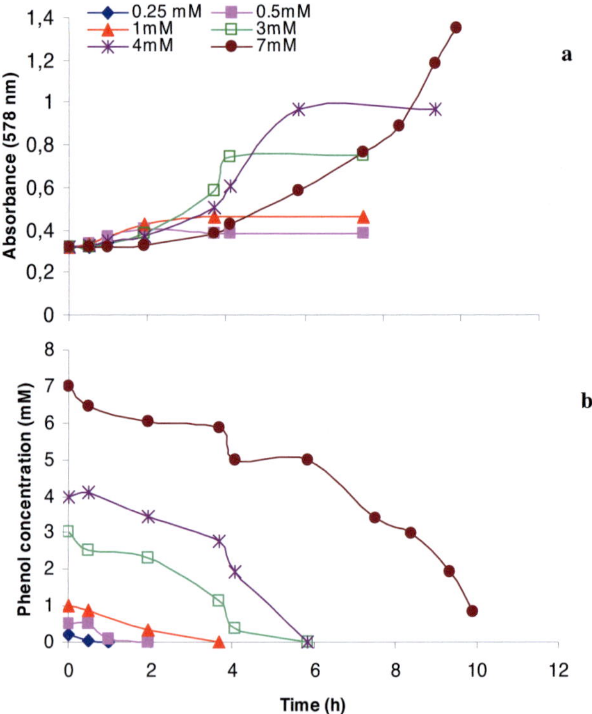

Fig. 3.35 - Time profile of different parameters: a) Growth of microbes as measured by optical density, b) Time course of phenol concentrations

Monod's model - one of the most popular growth model based on Michaelis-Menton enzyme equation (equation 3.11) is usually used to calculate the kinetic parameters for growth.

$$V = \frac{V_{max} \cdot S}{(K_s + S)}$$ (Equation 3.11)

In the Michaelis-Menton equation (Equation 3.11), V is the velocity of the reaction (mmol/l·h), V_{max} is the maximum velocity of the enzyme catalyzed reaction (mmol/l·h), K_s is the half saturation constant (mmol/l), S is substrate concentration (mmol/l).

In the Monod's equation (Equation 3.12), V and V_{max} are replaced by μ and μ_{max}

$$\mu = \frac{\mu_{max} \cdot S}{(K_s + S)}$$ (Equation 3.12)

μ is the growth rate (h) and μ_{max} is the maximum growth rate (h).

This model is in particular good for those substrates which are non-inhibitory substances for growth of bacteria. A modified version of this model is the Haldane equation (equation 3.13), which incorporates the inhibitory effect of toxic substrates

$$\mu = \frac{\mu_{max} \cdot S}{[K_s + S + (\frac{S^2}{K_i})]}$$ (Equation 3.13)

Where K_i is the inhibition constant (mmol/l).

The microbial flora adapted to the inhibitory substance tends to follow Monod's growth equation as μ increases initially with increase in the substrate but at a certain concentration it tends to decrease due to the toxic effect exerted by the substrate. Haldane model perfectly describes the phenomena of the combined effect of S as a growth substrate and as an inhibitor. Therefore this model was used to calculate the kinetic parameters in the present study because the microbial flora, although adapted to phenol, was somewhat inhibited at higher phenol concentrations (Fig. 3.34). The specific growth rate (μ) for each concentration was calculated from the slope of linear logarithmic plots of optical density against time (Moneteiro et al., 2000). Figure 3.35 is the semi-log graph used to calculate μ from the curves of optical density as a function of time.

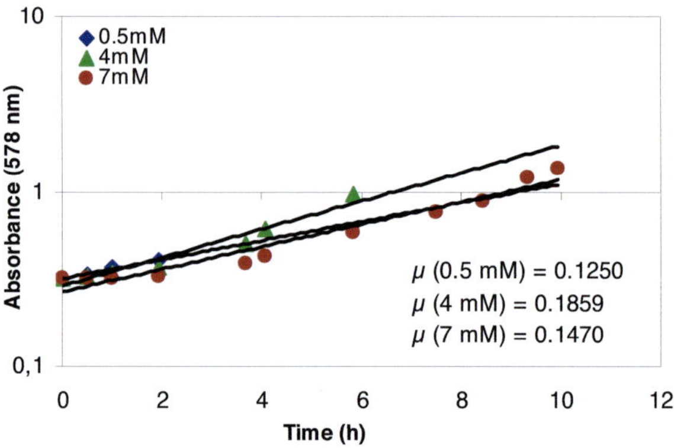

Fig. 3.36 –Growth rate determination for some of the test concentrations of phenol

The inhibition constants K_i and half saturation constant K_s, also defined as the substrate concentration at which μ is equal to half μ_{max}, in the Haldane equation were obtained using the nonlinear regression and fitting the experimental data of the specific growth rate as a function of phenol concentration by application of the GraphPad software- Prism 5.

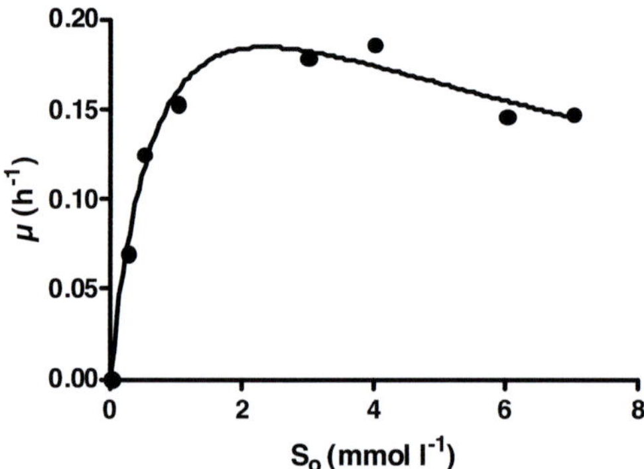

Fig. 3.37 - Determination of μ_{max} by non-linear regression according to Haldane model

The curve obtained (Fig. 3.36), had a high correlation coefficient (r^2) of 0.986. It indicated that the maximal growth rate of the mixed phenol-degrading enrichment was obtained at 2 mmol/l phenol. Below this concentration growth seemed to be suboptimal due to substrate limitation and above this concentration growth was inhibited increasingly due to substrate inhibition. The best fit values of the parameters are presented in Table 3.8.

Table 3.8: Growth kinetic parameters of the Haldane model for biodegradation of phenol by mixed culture, derived from the experimental data

Parameters	μ_{max} (h^{-1})	K_s (mmol/l)	K_i (mmol/l)
Best fit values	0.3095	0.7933	6.887
Std. error	0.04993	0.2304	2.519
95% Confidence intervals	0.181 to 0.438	0.201 to 1.386	0.410 to 13.36

The yield factor (Y) of the biomass was calculated as

$$Y = -\frac{dX}{dS}$$
(Equation 3.14)

Where dX is the change in biomass in terms of optical density related to change in the substrate concentration dS (Table 3.9). The decrease of substrate concentration first becomes notable when sufficient microbes are present, and the process stops when all the substrate has been consumed, therefore the formula for yield is usually represented with a negation. X was replaced with the OD at 578 nm, assuming a linear relationship between X and OD for OD-values below 0.7. The yield factor Y obtained for different concentrations of phenol (Table 3.9) was similar for all concentrations, except for the lowest concentration of 0.25 mmol/l and was in the range of 0.143-0.166.

Table 3.9: The yield factor (Y) of biomass calculated for different concentrations

Concentration (mmol/l)	Yield factor (A$_{578}$ biomass mg/l)
0.25	0.100
0.50	0.164
1	(0.143)
3	0.161
4	0.163
7	0.166

3.2.2 Anaerobic batch assays

A number of anaerobic batch assays for phenol degradation were conducted from time to time with inoculum sludge taken out from either the suspended or the fixed bed reactor. Some of the results are presented in this section.

3.2.2.1 Phenol degradation profile with sludge from the suspended bed reactor

A batch assay was carried out in serum bottles with the sludge from the anaerobic suspended bed reactor taken out on day 143 of reactor operation. The working volume of the assays was 50 ml. Phenol was added as the carbon source at concentrations of 2, 1 and 0.5 mmol/l and the assays were monitored for phenol degradation, SAC reduction and methane formation. The results are presented in Figure 3.37. At all the concentrations degradation of phenol was accompanied with a reduction of SAC, indicating degradation of the aromatic ring. The methane production was equal to the expected values as calculated by 'Buswell Equation' (equation 3.1).

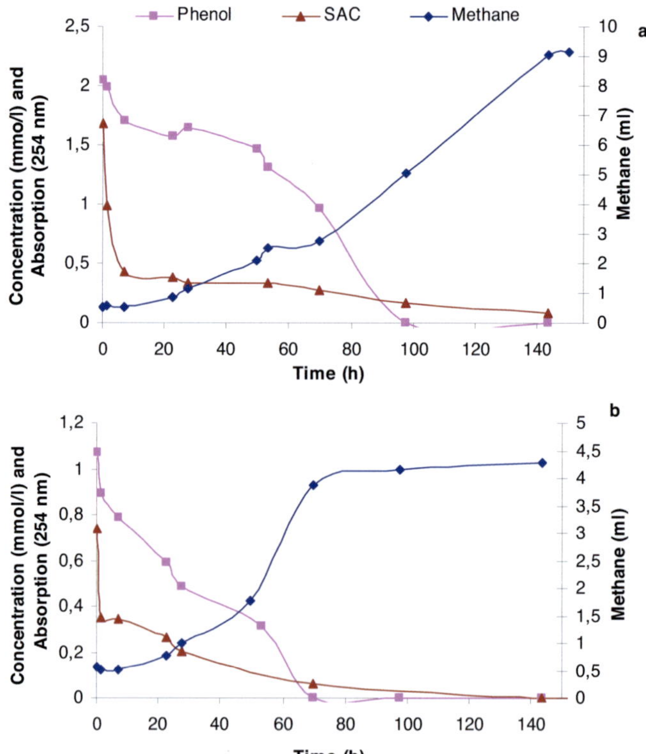

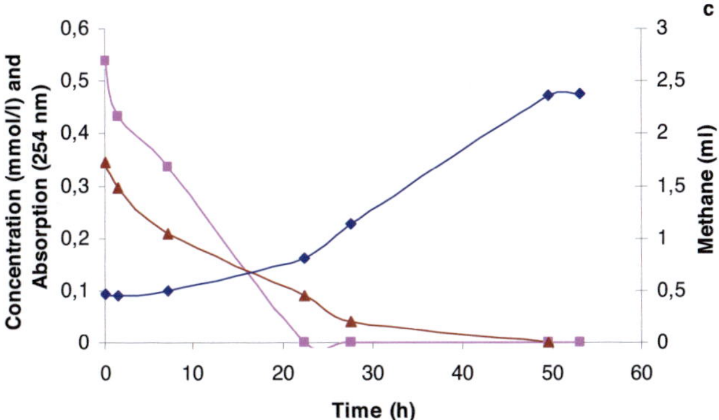

Fig. 3.38 – Phenol degradation, SAC reduction and methane production in assays with different phenol concentrations (a = 2 mmol/l, b = 1.1 mmol/l, C = 0.55 mmol/l)

1 mmol phenol = 94.1 mg

The maximum rate of phenol degradation observed was 3.12 mmol/l·d (294 mg/l·d), in the assay with 1 mmol/l phenol concentration, whereas the maximum degradation rate in assay with 2 mmol/l phenol concentration was only 1.2 mmol/l·d (113 mg/l·d). With the same inoculum, a batch assay was performed with benzoic acid (2 mmol/l) as the carbon substrate to demonstrate the ability of the culture for methane production. The relationship between decrease in SAC and methane production is represented in Figure 3.38. The complete expected methane production from the degradation of 2 mmol/l benzoic acid took less time than degradation of 2 mmol/l phenol, indicating that during phenol degradation, benzoic acid would not be accumulated, if benzoic acid would be an intermediate (Fig. 3.37).

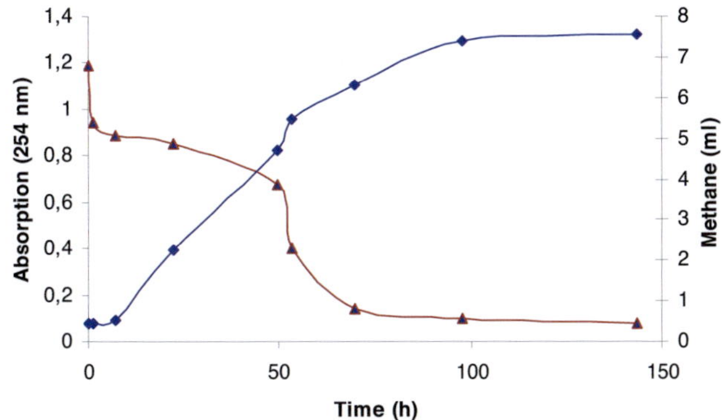

Fig. 3.39 – SAC reduction and methane production with benzoic acid as substrate

SAC was measured as absorption at 254 nm and represented the benzoic acid concentration

3.2.2.2 Phenol degradation profile with inoculum from the fixed bed reactor

It was difficult to take out the sludge immobilised on the carrier materials, therefore a batch assay was performed in serum bottles (116-119 ml) with only the effluent of the anaerobic fixed bed reactor collected over a period of 195 days of the reactor operation. The working volume of the assay was 50 ml to which different concentrations of phenol (1, 2, 3, and 5 mmol/l) were added. Phenol degradation and methane production are presented in the Figure 3.39. The maximum rate of phenol degradation was 10.2 mmol/l·d or 966 mg/l·d in an assay containing 2 mmol/l phenol. In all the assays, methane was produced after a lag phase, but the observed methane production matched to the expected theoretical values.

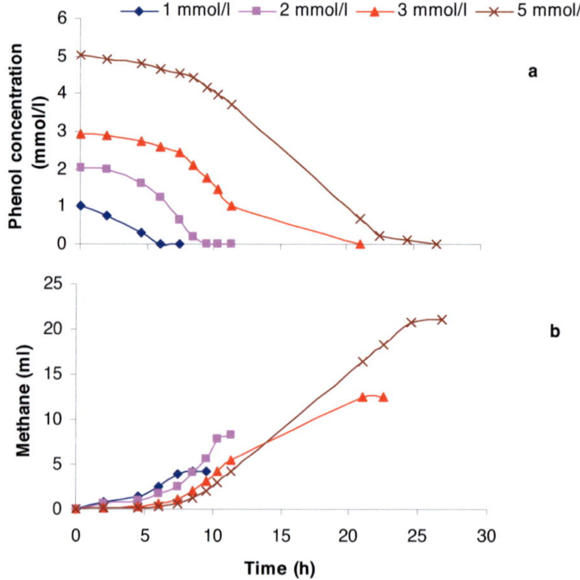

Fig. 3.40 – Phenol degradation profile with the anaerobic fixed bed inoculum in a batch assay

a) Phenol removal, b) Methane production

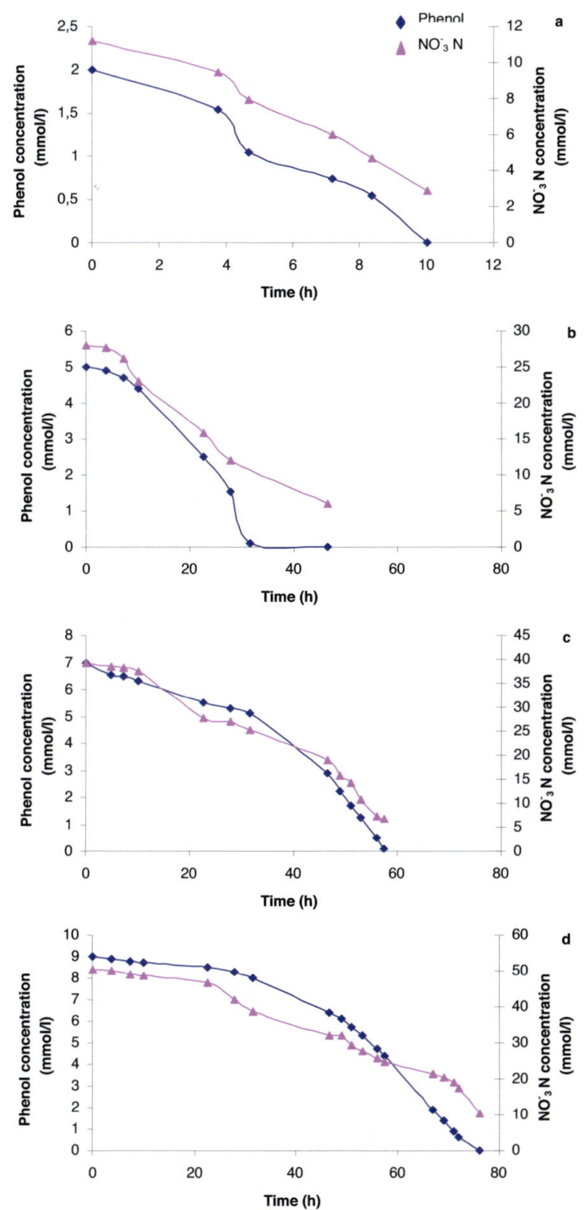

Fig. 3.41 –Denitrification in batch assays at phenol concentrations of

2 mmol/l (a), 5 mmol/l (b), 7 mmol/l (c) and 9 mmol/l (d)

1 mmol phenol = 94.1 mg

3.2.3 Anoxic batch assays

3.2.3.1 Phenol degradation profile under anoxic conditions

These assays were performed in serum bottles (116-119 ml). The working volume for the assays was 50 ml. The seed inoculum was 5 % sludge taken out from the suspended anoxic reactor on day 690 of the reactor operation. For denitrification, 565.6 mg/l KNO_3 was added per 1 mmol/l (94.1 g/l) phenol concentration which corresponded to required 78.32 mg/l NO_3^--N for the degradation of 1 mmol/l phenol or 5.6 mmol of NO_3^--N per 1 mmol of phenol (see equation 3.4). The calculated amounts of KNO_3 were added to 50 ml assays to reach the respective phenol concentrations.

Phenol was degraded at a maximum rate of 0.53 mmol/l·h (1.2 g/l·d) in the assay with 2 mmol/l phenol concentration. The theoretical ratio of phenol: NO_3^--N was 0.18. Whereas the observed values in all the assays (Table 3.10) were higher because some of the carbon was required for cell growth and maintenance, requiring less nitrate theoretically required for the phenol biodegradation. During the degradation no significant nitrite as well as accumulation of fatty acids as intermediatery metabolites was observed.

Table 3.10 : Degradation rates of phenol at different initial concentrations

Phenol Concentration (mmol/l)	2	5	7	9
Phenol : NO_3^--N	0.24	0.23	0.21	0.23
Phenol degradation rate (mmol/l·d)	12.8	9.3	6.7	7.2

3.3 Biodegradation of 2-CP in an anaerobic fixed bed reactor

An anaerobic fixed bed reactor (AFBR) was started by introducing the supernatant of a previously 2-CP-acclimatised sewage sludge under batch conditions to achieve efficient inoculum for the fast degradation of 2-CP under continuous cultivation conditions.

3.3.1 Preparation of inoculum - Acclimatisation of sewage sludge to 2-CP in a batch reactor

The time required for the degradation of 50 mg 2-CP/l (0.39 mmol/l) by non-acclimatised sewage sludge was 23 days. The degradation time was improved by 9 days upon the second addition of the same amount of 2-CP. When the sludge was spiked a third time with 2-CP, removal of 50 mg/l of 2-

CP took only 10 days. Up to 10 mg phenol/l accumulated when the 2-CP concentration was diminished to about 25 mg/l during the phase of rapid 2-CP degradation and the phenol disappeared again when the 2-CP concentration was below 20 mg/l (Fig. 3.41). Quick successive feeding of 2-CP upon complete substrate depletion reduced significantly the lag time before degradation started (Fig.3.41).

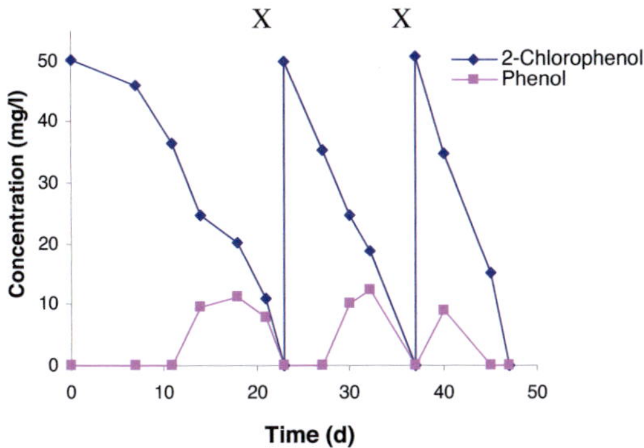

Fig. 3.42 – Acclimatisation of anaerobic sewage sludge to 2-chlorophenol
2-CP addition at the time marked with X

3.3.2 Acclimatisation of the microbial flora to increased loading rates of 2-CP in the anaerobic fixed bed reactor

The 1200 ml continuous anaerobic fixed bed reactor was started by introducing the supernatant of sludge acclimatised to 2-CP under batch conditions. Synthetic wastewater with 250 mg/l 2-CP was pumped into the reactor to maintain a HRT of 5 days and an OLR of 0.23 g COD/l·d. Growth of a biofilm on the clay beads was visible within 2 months of reactor operation. Complete mineralization of 2-CP starts with reductive dechlorination, anaerobic phenol degradation and production of biogas (equation 3.15).

$$C_6H_4OHCl + 4.5\ H_2O \longrightarrow 3.25\ CH_4 + 2.75\ CO_2 + HCl \qquad \text{(Equation 3.15)}$$
$$(128.5\ g \longrightarrow 6 \times 22.4 = 134.4\ l\ biogas)$$

The pH in the reactor was that of the synthetic wastewater (pH 7.3), indicating that the 50 mM phosphate buffer in the medium had sufficient buffer capacity. Therefore pH related problems, as often encountered in anaerobic biodegradation experiments were not experienced throughout the study period. Figure 3.42 represents 2-CP concentration in influent and effluent and % 2-CP removal when 2-CP was the main carbon source. No phenol accumulated, even in phases where the 2-CP was not completely degraded due to overloading.

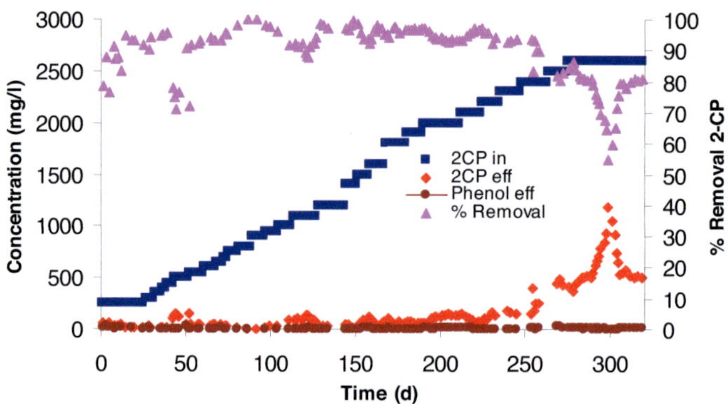

Fig. 3.43 – Reactor efficiency of 2-chlorophenol removal
1 mmol 2-CP = 128.56 mg

Phenol is the first intermediate product of reductive 2-CP dechlorination. Theoretically 1 mole of 2-CP should yield 1 mole of phenol. During batch acclimatisation of sewage sludge to 2-CP, phenol accumulation for a short time was observed (Fig. 3.40). This may be due to a higher rate of dechlorination of 2-CP compared to the rate of phenol degradation at surplus substrate availability during the acclimatisation phase. However, no or only traces of phenol were detected during continuous reactor operation, indicating that under balanced fermentation conditions phenol was apparently readily metabolized, either by the 2-CP dechlorinating bacteria themselves or by phenol-degrading bacteria in the sludge. The 2-CP removal during continuous fermentation in the reactor was 78 % of 0.25 g 2-CP/l, already after the first day. This high removal efficiency was only possible after a preceding acclimatisation of the sludge to 2-CP as the sole carbon source. In less than one month the steady removal efficiency for 2-CP reached > 90 % at a loading rate of around 50 mg/l·d. Within 278 days the concentration of 2-CP in the feed medium was increased stepwise up to 2.6 g/l (20.2 mmol/l) to reach a 2-CP loading rate of 760 mg/l·d (6 mmol/l·d) at a HRT of 3.4 days (Table 3.11). The average 2-CP removal at this 2-CP loading rate was 85 %.

Table 3.11: Reactor conditions during 318 days of operation with 2-CP

Days	Influent 2-CP concentration mg/l	Average HRT d	2-CP loading rate mg /l·d
1-25	250	4.9	51
26-31	300	5.4	56
32-35	350	4.7	75
36-38	400	5.0	80
39-42	450	5.0	90
43-51	500	5.0	100
52-60	550	4.8	115
61-67	600	4.5	133
68-71	650	4.7	104
72-74	700	4.6	154
75-80	750	4.6	165
81-88	800	4.6	173
89-96	900	4.4	204
97-104	950	4.7	204
105-112	1000	4.8	208
113-127	1100	4.5	248
128-143	1200	4.3	280
144-150	1400	4.3	328
151-157	1500	4.3	350
158-167	1600	4.3	373
168-179	1800	3.6	496
180-189	1900	3.6	530
190-211	2000	3.6	558
212-223	2100	3.6	588
224-234	2200	3.5	623
235-246	2300	3.5	652
247-262	2400	3.5	693
263-273	2500	3.4	729
274-278	2600	3.4	758
279-288	2600	2.7	975
289-296	2600	2.2	1192
297-298	2600	1.9	1365
299-300	0	0	0
301-318	2600	3.4	758

From day 279 onwards the 2-CP concentration was kept at 2.6 g/l and the HRT was successively reduced from 3.4 to 1.9 days to increase the 2-CP-loading finally to 1365 mg/l·d (Table 3.11). The 2-CP loading rate during this period (day 279-298) ranged from 0.76 g 2-CP/l·d to maximally 1.35 g 2-CP/l·d. The highest 2-CP degradation rate in the reactor, of 0.87 g/l·d was observed during days 289-296 at a HRT of 2.2 days for a 2-CP loading rate of 1.19 g/l·d and an influent 2-CP concentration of 2.6 g/l (Table 3.11).

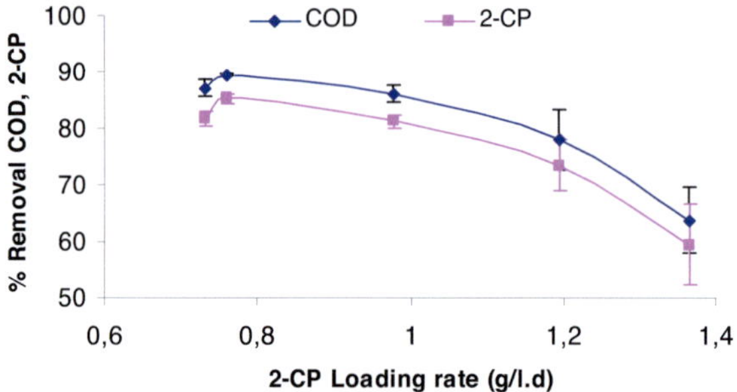

Fig. 3.44- Response to increased 2-CP loading rates in an AFBR

The increase of the 2-CP loading rate from 0.69 to 1.36 g/l·d led to a constant drop in the 2-CP and COD removal. A more rapid decrease of removal efficiencies for COD and 2-CP was observed when the loading rate exceeded 1 g 2-CP/l·d (Fig. 3.42). At a loading rate of 1.36 g 2-CP/l·d (equivalent to an OLR 2.67 g COD/l·d), the 2-CP removal rate fell to 54 %. To stop the decreasing removal efficiency and to allow recovery of the reactor flora, no feed was provided for the next 2 days. Then continuous feed addition was re-established to maintain a HRT of 3.4 days and a 2-CP loading of 0.73 g/l·d. Within 2 weeks the 2-CP removal rate recovered and reached 80 % (Fig. 3.41, from 300 days onwards). This indicated that, although the reactor was sensitive towards high loadings of 2-CP, it was capable of sustaining a short period of toxic overloading, provided that the overload condition was reverted quickly to tolerable loading rates.

3.3.3 COD removal during acclimatisation of the microbial flora to high 2-CP concentrations in the anaerobic fixed bed reactor

The performance of COD removal during the reactor operation is shown in Figure 3.43. From equation 3.16, a theoretical chemical oxygen demand of 1.62 g per 1 g 2-CP can be derived.

$$C_6H_4OHCl + 6.5\ O_2 \longrightarrow 6\ CO_2 + 2\ H_2O + HCl \qquad \text{(Equation 3.16)}$$

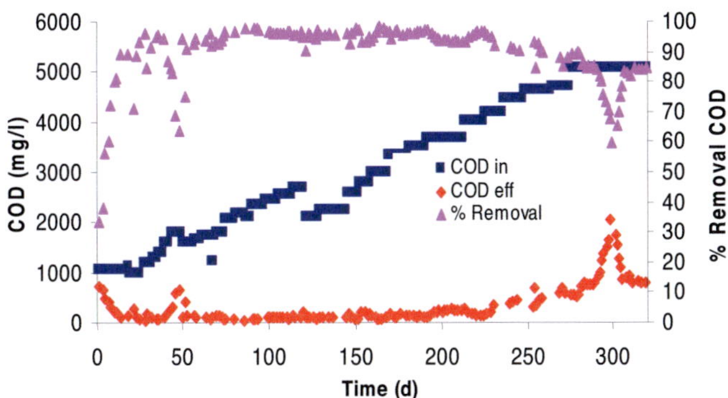

Fig. 3.45 – COD removal in the AFBR fed with a 2-CP containing medium

At the start of the reactor, the total influent COD was about 1 g/l, whereas the total influent 2-CP concentration was 250 mg/l, corresponding to about 405 mg/l COD in the medium. About 595 mg/l COD in the influent was due to yeast extract and peptone in the feed. The COD removal was not different from the 2-CP removal in the reactor. At the start 78 % of 2-CP (195 mg/l) and only a total of 330 mg/l of 1000 mg/l COD present was removed. The COD removal was thus mainly due to 2-CP degradation, indicating that the microbial flora, which was well adapted to 2-CP, had an affinity for 2-CP. This may be due to the fact that the reactor seed was acclimatised to 2-CP as the only carbon source during pre-adaptation. Within the following 10 days, the COD removal increased to 80 %, indicating the ability of microbes to also utilise other substrates such as yeast extract or peptone of the feed.

On day 118 of reactor operation, yeast extract and peptone as semi-defined sources of nitrogen and vitamins in the feed were replaced by 0.5 g/l NH_4NO_3 and 1.3 ml/l of a vitamin solution. On day 120 COD and 2-CP removal both had dropped by about 5-3 %, but recovered to the original removal efficiencies within 3 (COD) or 7 (2-CP) days (Fig. 3.44). This indicated that organic nitrogen sources could be replaced by inorganic nitrogen sources with only a short negative effect on removal efficiencies.

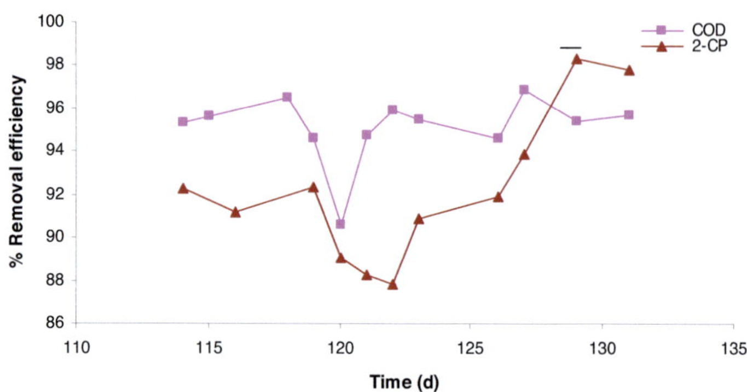

**Fig. 3.46 - Response of vitamin and NH_4NO_3 addition instead of peptone
and yeast extract in the reactor feed solution**

When the 2-CP loading rate was changed by decreasing the HRT on day 279 (Table 3.11), a drop in both, COD removal and 2-CP removal was seen (Figs. 3.41 and 3.43). At an OLR of 2.67 g COD/l·d and a 2-CP loading rate of 1.35 g/l·d, the removal of COD dropped to 70 % and of 2-CP to 54 %. When loading was interrupted for 2 days and then restarted as described in section 3.3.2, the removal efficiencies recovered within a week. At the end of reactor operation on day 318, the COD removal was 84 % and 2-CP removal was 81 %. Anaerobic degradation of 2-CP at an influent concentration as high as 2.67 g 2-CP/l has not been reported before.

3.3.4 Chloride ion, biogas and fatty acids production

The removal of 2-CP starts with dehalogenation of chlorophenol, accompanied by the production of inorganic chloride ions. It can be deducted from the reaction stoichiometry that during dechlorination of 1 g 2-CP, 0.28 g of inorganic chloride must be released. The inoculum used for the reactor start-up already contained chloride. Therefore the concentration of chloride was higher than expected from complete dechlorination after start-up (Fig. 3.55). At a HRT of 5 days the concentration of the chloride was diluted within a week to a concentration that could stem from 2-CP dechlorination. The ratio of chloride released per unit of 2-CP removed was then close to the theoretical value (Fig. 3.56).

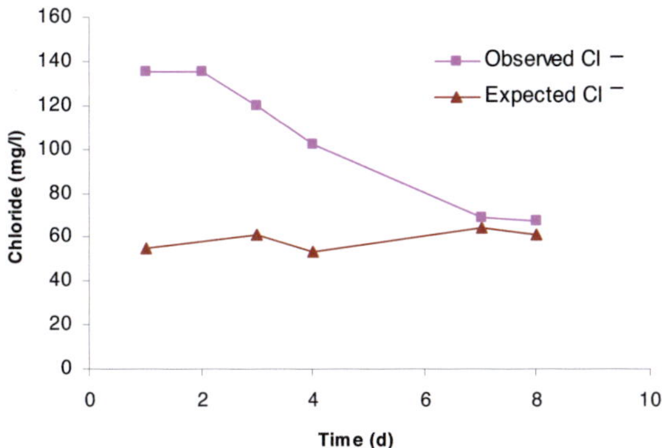

Fig. 3.47 - Observed and expected chloride ion production at reactor start-up

The observed range from day 8 to day 317 of reactor operation was 0.18 to 0.31 with 0.20 as the average ratio and 0.24 as the median value, indicating some fluctuations in dehalogenation of 2-CP during the study period.

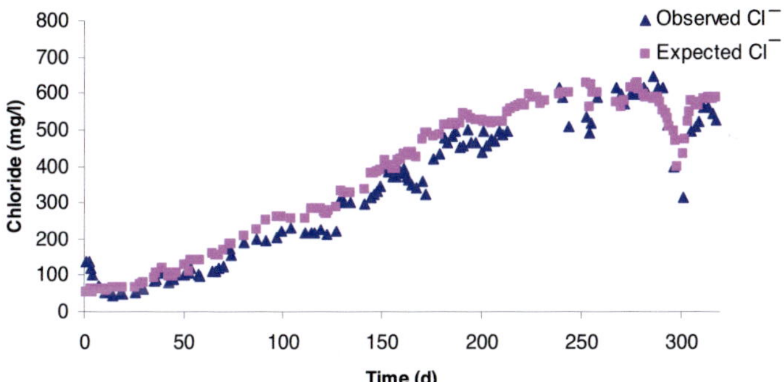

Fig. 3.48 - Observed and expected chloride ion production throughout reactor operation

During 2-CP degradation, the accumulation of volatile fatty acids in the reactor was negligible. The maximum fatty acid concentration was observed in the reactor effluent on day 298 with 144 mg/l acetic acid and 37 mg/l propionic acid, when 2-CP loading reached its peak value. Less fatty acids were an indication of rapid acetogenesis and methanogenesis during steady state conditions or of an inhibition of acidogenesis by 2-CP at higher loadings when only 54 % of 2-CP were removed. The theoretical biogas production from 2-CP could be calculated from equation 3.16. From 1 g 2-CP 1.05 l biogas could be obtained at standard temperature and pressure conditions. The 2-CP loading rate during the study period ranged from 0.05 g/l·d to maximally 1.3 g/l·d (only for 2 days). Therefore the amount of biogas expected for complete methanogenesis was very low. The observed amount of biogas ranged from 0 to maximally 0.55/l·d. The biogas consisted of 54 – 67 % CH_4.

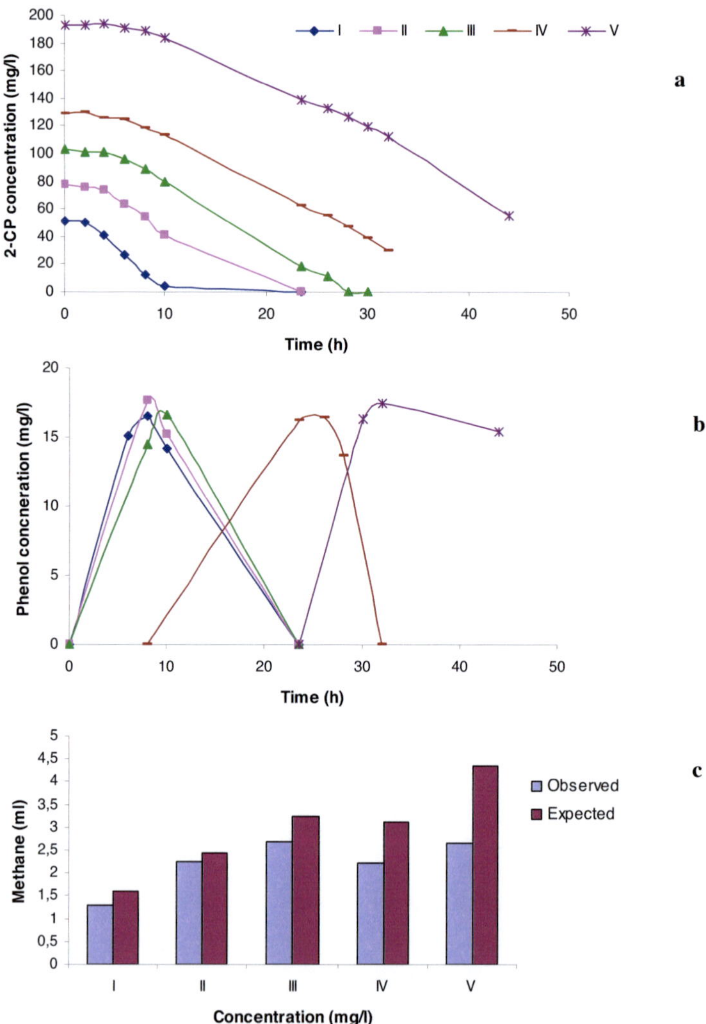

Fig. 3.49 - Batch assay a) Degradation of 2-CP b) Production of phenol and c) Observed and expected methane production

Assay I - 51.5 mg/l, II- 77 mg/l, III – 103 mg/l, IV-128.5 mg/l, V- 192.4 mg/l

3.3.5 Degradation of 2-CP under batch conditions

An assay was carried out to determine the rate of 2-CP degradation in a batch assay by the well acclimatised reactor flora (Fig. 3.57). Since the flora in the reactor was very sensitive to oxygen and it was not feasible to take out reactor sludge without the intrusion of oxygen in the reactor, reactor effluent was collected anaerobically during days 149 –157 and used for the experiment. Even after gravity sedimentation of biomass the volatile suspended solids content was only 2.3 g/l. During the batch assays 13.7 to 17.6 mg/l phenol was intermediarily produced from 50 – 190 mg/l 2-CP (Fig. 3.57 b). The mineralization of 2-CP was triggered at all test concentrations as evident from the production of methane (Fig. 3.57 c). The highest rate of 2-CP degradation was 175.2 mg/l·d which was observed at 50 mg/l 2-CP concentration, whereas at 192 mg/l, the highest degradation rate was only 114.7 mg/l·d. The 2-CP concentration of 500 mg/l didn't show any degradation even after 120 hours of incubation (not shown in the figure).

3.3.6 Effect of Co-substrates on 2-CP degradation

The effluent from the reactor was collected again on days 212-217, to carry out a batch test of 2-CP degradation in the presence of co-substrates such as peptone, yeast or glucose. The degradation of 2-CP and the production of methane started after a lag phase, except for the assays where glucose was a co-substrate (Fig. 3.3.6). The presence of peptone and yeast, respectively, had no effect on the 2-CP degradation, but when both where present together, only a slight decrease in degradation was observed. However, in assays which contained glucose, a significant decrease in biodegradation of 2-CP was observed as only 59 % of 2-CP could be degraded after 29 hours of incubation compared to > 80 % in all other assays. The expected methane from the degradation of 2-CP was 3.9 ml and only 2.7 ml could be observed after 41 hours in the control. From the glucose containing assay, the methane was measurable soon after 1.25 hours of incubation. In all assays no phenol accumulation was observed except in the assay with glucose (results not shown). The observed phenol concentration in this assay was up to 23 mg/l after 29 hours of incubation.

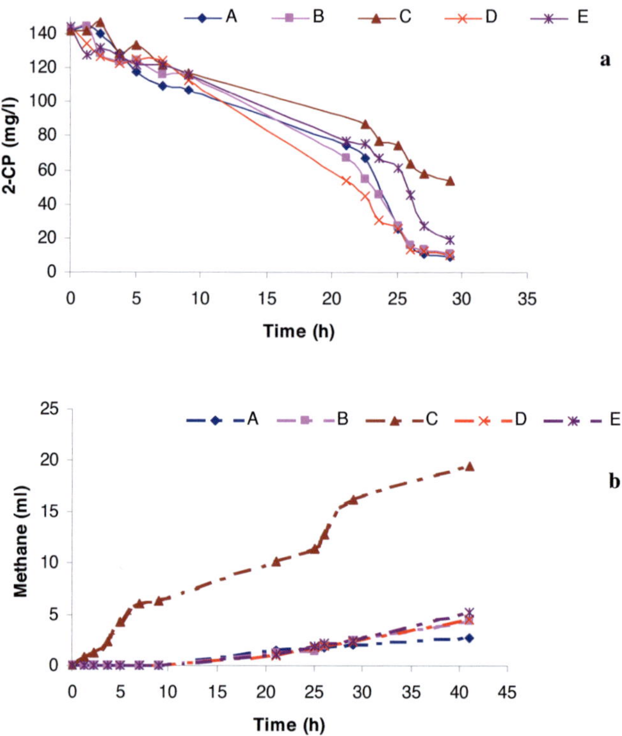

**Fig. 3.50 – Effect of co-substrate addition on 2-CP degradation (a) and 2-CP concentration
and methane production (b)**

All assays contained 140 mg/l 2-CP, whereas assay B, C, D and E additionally were supplemented
with 250 mg/l peptone (assay B), 1 g/l glucose (assay C), 250 mg/l yeast extract (assay D) and 250
mg/l peptone + 250 mg/l yeast extract (assay E), respectively.

Chapter 4

Discussions

4.1 Phenol biodegradation in suspended bed reactors

Municipal sewage sludge was used as an inoculum for the start-up of aerobic, anaerobic and anoxic suspended bed reactors, as domestic sewage sludge is the richest source of microbes. Glucose was utilised as the main carbon substrate for start-up of the reactors because it is a readily and completely biodegradable organic compound. Complete mineralization is a growth related process. Aerobic microbes responsible for converting the organic substrate into inorganic products, utilize most of the substrate for respiration with energy conservation for growth, accompanied by some release of energy as heat. Up to 50 % of the carbon are assimilated for biosynthesis with the conserved energy. This is associated with an increase of cell numbers and biomass (Richards and Shieh, 1986). Therefore it was desired to have adequate and active biomass in the reactor before phenol was introduced since phenol is a known growth inhibitory substance. For the reactor start-up, feeding the sewage sludge with the phenol would have resulted in partial degradation or complete mineralization (at very little phenol concentration) or no degradation at all, depending on the presence of phenol-degrading microorganisms in the inoculum sludge. Thus, toxicity might have prevented start-up if only a partial or no degradation of phenol had occurred or the phenol-degrading bacteria were under-represented and grew and utilized phenol only very slowly. Industrial wastewaters often contain high concentrations of COD. For laboratory investigations, glucose is a very good substrate to provide high COD concentrations (upto 5 g/l) in the reactors. By this less toxic industrial wastewater was simulated, which would have been extremely difficult, if using highly concentrated and thus toxic phenolic wastewaters at the start-up phase. Once it was ensured that reactors were capable to degrade high COD concentrations under different organic loading rates, phenol was introduced in all the reactors.

4.1.1 Aerobic phenol degradation

The utilization of easily biodegradable substrates such as glucose ensured quick growth in the system. However, after addition of phenol to the reactor, the reactor performance remained more or less unstable. The results (Fig.3.2) obtained in this study seem to suggest that aerobic sludge grown on glucose for long time periods was sensitive to the presence of an inhibitory organic compound such as phenol. At the time of phenol introduction, the reactor had a high bioactivity (Fig 3.1) which was thought to increase the volumetric conversion capacity and the flexibility to the fluctuated loading rates and to tolerance of growth inhibitory substances like phenol. However, the addition of 0.19 g/l phenol to the glucose-fed sludge culture inhibited the activity and consequently led to accumulation of the soluble organic matter in the reactor. A reduction in removal efficiencies after addition of phenol in acetate fed aerobic granules was also observed by Liu et al., 2002.

For aerobic biodegradation, oxygen (either supplied by aeration or by injection of pure oxygen) is essential for the respiration to produce the required energy for growth and maintenance. The accidental interruption of the air supply of the reactor over a weekend (Fig. 3.2) caused hindrance in the degradation activity, and only 44 % COD removal was observable since an oxygen concentration below 20 % saturation was strongly rate limiting for aerobic processes (Worden and Donaldson, 1987). Activated sludge of a sewage treatment plant was provided to the reactor to increase the reactor biomass and thus the degradation activity. This aerobic reactor was sensitive to change in conditions and required long stabilisation times. The decrease in the reactor efficiency was often accompanied with colour changes of the sludge from brown to pale yellow and with a decrease of biomass in the reactor. Zheng et al. (2006) also observed the change in colour of sludge from greyish brown into yellow for some time in a SBR. This colour change was due to the formation of yellow coloured small granules which infact had increased the settling ability of the black coloured filamentous sludge contrary to the present study where yellow sludge in the reactor had a thin appearance due to the dominance of yellow coloured irregular flocs which had poor settling abilities and caused sludge aggregation. This behaviour might be influenced by the reactor conditions over the time which had initiated the growth of such microbes and their dominance over the brown coloured flora.

The disconnection of feed for approximately one month was done in order to initiate the consumption of accumulated organic compounds and thereafter to starve the microbial flora. It has been demonstrated that starvation improves survival of bacteria introduced in the phenol acclimatised activated sludge (Watanabe et al., 2000). During starvation changes in cell physiology including changes in cell shape (Givskov et al., 1994 a; Van Overbeek et al., 1995), RNA content (Givskov et al., 1994 b), protein expression pattern (Givskov et al., 1994 b), motility (Wei and Bauer, 1998), amounts of extracellular polymers (Wrangstadh et al., 1990), and stress resistance (Van Overbeek et al.,1995) have been observed. These starvation induced changes may help the better survivability of cells under adverse conditions (Watanabe at al., 2000). Starvation is also thought to play a crucial role in the aggregation of cells into granules (Tay et al. 2001 a, Tay et al., 2004), and thus reducing the susceptibility of cell washouts.

In the present study, the reactor performance was also improved after starvation. However the reactor was still sensitive for an increase in phenol concentration, indicating in this case that starvation was not the long term solution to achieve high efficiencies. The maximum phenol removal rate observed in this study was 546 mg/l·d at 753 mg/l concentration. The suspended or free cell systems had presented strong inhibitory effects for phenol concentrations as little as 100 mg/l (Pawlosky and howell, 1973; Dikshitulu et al., 1993).

4.1.2 Anaerobic phenol degradation

The enrichment of the mixed microbial population from the anaerobic sewage sludge was done by growing the flora in a synthetic medium, containing glucose. After ensuring a stable and high removal rate with glucose as carbon source, phenol was added in the reactor. To enable the culture adaptation to the toxic substrate phenol, the OLR was decreased. The capacity of anaerobic methanogenic bacteria to remove COD and phenol after extended carbon starvation was slightly effected as the flora was able to maintain its previous removal efficiencies within a week after reactor start-up. Roslev and King (1995) had reported that during anaerobic starvation, only the concentration of intracellular low-molecular-weight compounds decreased, and no significant changes were measured for cellular protein, lipids, polysaccharides, and nucleic acids compared to aerobic starvation. This might be the reason behind the more persistent capacity of anaerobic metabolism of phenol in bacteria after long term starvation. The problem of a lower volume of observed biogas than expected during anaerobic degradation was mainly due to the measurement inaccuracies. The gas meter used was best functional under a daily flow of at least 24 litre of gas whereas the expected gas volume was far less than that. The running of the reactor under batch conditions and measuring the gas volume with water displacement confirmed this.

Transient changes in operating parameters like accidental temperature drops affected the integrity of the anaerobic process and resulted in a drop in COD removal due to accumulation of fatty acids, mainly acetic acid, indicating that acidogenic bacteria were sensitive to the temperature drop. About 1 month was required for the improvement of COD removal to the previous efficiency. Cha et al. (1997), also observed a dramatic decrease in COD removal efficiency after a temperature drop, which was increased to a certain level after about 15 days, and finally was maintained at that level. This was mainly due to a decrease in acidogenesis, which eventually produces acetate, serving as a starting substrate to produce approximate 70% of methane. However, the number of acetogenic bacteria remained unaffected due to this temperature drop. In the present study, the phenol removal was gradually affected compared to COD removal, but after some time the removal of phenol dropped down considerable. This may have been due to conversion of phenol to benzoate- a common intermediate of phenol metabolism -, which was not detectable by the gas chromatography method applied. This view is supported by the study of Leven et al. (2006), who reported benzoate accumulation during anaerobic phenol degradation, when the incubation temperature was changed, likely due to disturbance of the methanogens, or activation of another degradation pathway or organism. Disturbed methanogenic activity can cause accumulation of acetate and hydrogen, with a subsequent increase in benzoate (Knoll and Winter, 1989; Kobayashi et al., 1989; Karlsson et al., 2000). Under these conditions, the conversion of phenol to benzoate is initially not affected, but if benzoate accumulates to high concentrations, conversion of phenol is also inhibited (Knoll and Winter, 1987; Wang and Barlaz, 1998). Transformation of benzoate recovers when levels of acetate and hydrogen eventually decrease (Knoll and Winter, 1989).

The performance of the reactor after a second starvation period was more or less the same, indicating that starvation might not effect the efficiency during shut down periods for e.g. maintenance or repairs. The optimum phenol loading for this reactor was 0.48 g/l·d with a phenol removal rate of 0.44 g/l·d at an influent concentration of 1.13 g/l, indicating the suitability of this kind of suspended bed reactor for phenol loading rates of ≤ 0.5 g/l·d.

4.1.3 Anoxic phenol degradation

Glucose was used as the initial substrate for the enrichment of this suspended bed reactor too (see Sections 4.1.1 and 4.1.2). In the anoxic environment there are several possible fates of NO_3, including: assimilatory nitrate reduction (immobilization), dissimilatory nitrate reduction to ammonia and denitrification (Tiedje, 1988; Greenan et al., 2006). The terminal end product of denitrification is N_2 gas. If substantial amounts of NO_3 were converted to NH_4 or organic N rather than N_2, then N may be exported from the system in that form, or remain in the system and later be converted to NO_3 thus defeating the treatment strategy. This bioreactor has demonstrated good nitrate removal from synthetic wastewater containing glucose as the carbon source (Fig. 3.16). The gas was analysed from time to time and no or only traces of CH_4 were found, indicating successful denitrification. After addition of phenol to the reactor the removal was satisfactory and an organic shock load by a factor of 2 or a more drastic phenol shock were not harmful, if the previous reactor conditions were maintained. A similar observation was made by Tay et al. (2001 b), who studied phenol shock loading in UASB reactors and stated that reactors were able to recover fully when the shock loading of phenol was returned to the pre-shock level. Uygur and Kargi (2004) reported a decrease in removal from 95 % to 79 % of 1200 mg/l COD when phenol concentrations were increased from 400 mg/l to 600 mg/l in a four step (anaerobic/oxic/anoxic/oxic) sequencing batch reactor. The COD concentration in their synthetic wastewater was < 4 times the influent COD concentration of the present study. Nevertheless one can observe the inhibition effects of phenol on COD removal at concentrations around 400 mg/l.

In the current study glucose monohydrate, a readily biodegradable substrate was provided along with phenol as substrate. The increase in phenol (day 450, Fig. 3.18), caused a drop in the COD as well as phenol removal, indicating that phenol had a severe toxic effect on those bacteria which could readily degrade glucose. According to the oxidation stoichiometry, COD of 1 g glucose monohydrate is 0.969 g and COD of 1 g phenol is 2.38 g. Taking data of day 450 (Table 4.1), it could be deducted that the majority of COD in the effluent was from residual glucose rather than phenol.

Table 4.1 : COD contribution from glucose and phenol in reactor effluent on day 450

COD_{in} (g/l)	COD_{out} (g/l) X	$Phenol_{out}$ (g/l)	Phenol COD_{out} (g/l) Y	COD_{out} (from glucose and nutrients) (g/l) X-Y
5.321	2.897	0.2192	0.521	2.376

As soon as only the phenol loading was lowered, it attained the previous values in less than 9 days. This is due to subsequent lower toxicity exerted by phenol on microbial population. Chakraborty and Veeramani (2005) also analysed the effect of phenol shock load injection in an anaerobic, anoxic and aerobic suspended growth continuous system. The shock load injection was applied to the anaerobic reactor; the effluent from this reactor was influent for the anoxic reactor and contained 1.4 times more phenol (620 mg/l) than the usual loading. Their anoxic reactor also achieved its previous phenol removal efficiency in 7-9 days.

C/N ratios varying from 3.3 to 6 have been reported in the literature (Fang and Zhou, 1999; Sarfaraz et al. 2004, Eiroa et al. 2005) for denitrification. Anaerobic environments with a high C to electron acceptor ratio tend to favour dissimilatory nitrate reduction to ammonia over denitrification (Tiedje et al., 1982). However, precise information on critical C to electron acceptor ratios that control partitioning between these two processes is not yet available. In the present reactor this ratio was 3.5, rather close to the theoretical values (Section 3.1.1.3.3). The reactor performance was satisfactory at OLRs between 4-6 g COD/l·d, and over a long time period of reactor operation, it was capable to sustain a high OLR (> 5 to $\leq$ 10 g COD/l·d). Under unfavourable conditions more adverse effects on phenol removal than on COD removal were observable. In such conditions a polishing reactor to remove remaining COD and phenol would be complementary to such process.

4.2 Phenol biodegradation in fixed bed reactors

The effluents from the suspended bed reactor were used as inoculum to start aerobic and anaerobic fixed bed reactors, respectively. The phenol was provided in the influent during start-up of these reactors because the inoculum was already acclimatised to phenol. The effluent rather than sludge is a more suitable inoculum for the start-up phase of the fixed bed reactor as it prevents clogging and channelling in the reactor and ensures good settling of microbial flora on fresh carrier material.

4.2.1 Aerobic phenol degradation

This reactor was started to overcome the problems encountered in the suspended bed reactor, mainly acclimatisation of bacteria to phenol and subsequently to investigate phenol biodegradation at increasing influent concentrations from 4.7 to 5.2 g/l. A second aim was to analyse the reaction of the system for sudden changes in influent conditions, as often experienced in industrial treatment systems. The minimal nitrogen demand was elucidated, as nitrogen is the most important feed component after carbon. The stability of COD removal in this reactor depended upon the applied OLR, as COD removal efficiency was decreased when OLR was increased. In this study phenolic wastewater with phenol concentrations of up to 5.2 g/l was used. Most of the previous studies have used maximum influent phenol concentration between 0.5 to 3 g/l (Jiang et al., 2002; Shetty et al., 2007; Tay et al., 2005; Nuhoglu and Yalcin, 2005; Esparza et al., 2006; Marrot et al., 2006). However, phenol concentrations up to 10 g/l have been reported in industrial wastewaters (Fedorak and Hrudey, 1988). Such highly concentrated wastewater is not generally suitable for biological treatment without pre-treatment because microbial growth and concomitant biodegradation of phenol are hindered by the toxicity of high phenol concentrations (Tay et al., 2005).

Marrot et al. (2006) reported successful treatment of phenol concentrations up to 1 g/l in an immersed membrane bioreactor. They had used activated sludge from a wastewater treatment plant, grown in a continuous culture using phenol as the limiting substrate. However by using a pure strain of *Pseudomonas putida* ATCC 17484 in a fluidised bed (immobilized cultures), > 90 % removal of 4 g /l·d has been reported by Gonzalez et al., 2001. In their study, the phenol concentration was maintained at ca. 1 g/l in the feed, while the HRT was progressively decreased from 4 to 0.2 days. The system was stable up to a HRT of 0.25 days. However the bioreactor became unstable at a HRT of 0.2 days and the phenol concentration in the effluent increased progressively with washout of biomass. For an industrial wastewater treatment a mixed consortium is more preferable than pure cultures. Therefore mixed consortia were used for the aerobic fixed bed reactor. Esparza et al., (2006) evaluated aerobic phenol removal by biomass, entrapped in polymeric hydrogel beads of calcium alginate (1 %) and cross linked poly (N-vinyl pyrrolidone), x-PVP (4%). They tested three different media with various F/M ratios. Highest phenol removal between 67.5 and 125 mg/l·h at the same HRT but with different F/M, was observed in a medium, rich in phosphate and ammonium ions, whereas removal was lower with the other two media. It was also stated that phenol was adsorbed by hydrogel beads at an equilibrium adsorption of 0.273 $\pm$ 0.05 mg phenol/bead.

Moreover strong inconvenience with this carrier material and above said influent medium was experienced as hydrogel beads were deteriorated structurally by friction and only after 7 days of continuous treatment, the beads were destroyed completely. The saturated carrier medium used in present reactor was unable to adsorb phenol from the mineral solution when kept soaked for 48 hours (result not shown). Therefore adsorption is ruled out in present study.

Nitrogenous co-substrates have an important effect on biodegradation of toxic substances like phenol. Peptone, meat extract and urea served as main nitrogen sources for the biomass. In the absence of these substrates, a decrease in removal efficiencies was observed. In order to avoid the negative effect on the microbial flora, previous conditions in terms of nitrogen supply were maintained. However glucose dosage, which was also decreased from 5 to 3 g/l, was not resumed. This indicated that the microbial flora was capable of utilising toxic and not easily degradable organic matter, when part of the glucose portion in feed was replaced by phenol. In the absence of a protein source (or nitrogen), the effect of phenol toxicity was more predominant but the reactor could still recover.

4.2.2 Anaerobic phenol degradation

Before the shock loading, efficient phenol degradation was achieved in the anaerobic fixed bed reactor, which confirmed the feasibility of this wastewater treatment approach. Volatile fatty acid accumulation at the start-up of the reactor was reversed as soon as the reactor attained the steady state. The performance was found to improve with the number of phenol amendments, thus showing that microbial adaptation can occur throughout the trial. The length of the acclimatisation phase decreased after each phenol amendment, as similarly reported by Collins et al. (2005). The COD and the phenol removal before the shock load were concurrent, indicating that phenol was really removed from the liquid and not converted to intermediary products. Biochemical inhibition was not observed when immobilised biomass was subjected to phenol concentration of up to 3.76 g/l and high phenol removal efficiencies were attainable at such a high-strength phenol containing influent. It could have been enhanced if shock load was not applied. The methane content of the biogas was very satisfactory; however the overall gas production was lower than expected from the 'Buswell equation'. The actual biogas amount however must be lower and can be calculated by including a correction factor for the degree of degradability, the pH, which influences CO_2 absorption and a 5-10 % discount for biomass formation (Gallert and Winter, 1999). In this study the only correction factor applied was the degradation degree as measured from the COD removal. At increasing pH, CO_2 is readily soluble in water. The reactor pH was maintained at 7.3 with buffered synthetic wastewater, which may have caused diminishment in total amount of expected biogas as some amount of CO_2 was absorbed and solubilised in the liquid. Moreover the gas meter may not indicate correctly the low amount of daily gas production which was in the range of its minimal sensitivity. All these reasons might have contributed to apparently less produced biogas than theoretically expected throughout the study period. After the shock loading, a decrease in the methane content (compared to the time before phenol shock load) was observed until the end of the experiment, indicating inhibition of some of the trophic groups involved in the methanogenic activity by phenol. An acute decrease in COD and phenol removal was observed along with accumulation of phenol during the consistent phenol shock load. Tay et al. (2001 b), reported that the phenol removal efficiency dropped to 14 % from 88 % within 30 days after increasing feed phenol concentrations from 1.2 g/l to 1.67 g/l. There was an increase in the production of intermediates, namely acetic acid and benzoate. The recovery was observed within 2 months after disturbance and restoration of the phenol concentration to 1.25 g/l. However in an another reactor with 1 g/l glucose supplement, recovery was faster by 1 month suggesting that inhibition was reversible in reactors with or without glucose supplement, even after extended period of toxic exposure. Contrary to that Gali et al. (2006), who reported reactor failure when the phenol concentration was increased from 1.1 g/l to 1.34 g/l in a reactor treating phenol and readily degradable sugar cane based molasses after a shock load period of 20 days. The maximum influent COD concentration was only 4 g/l. An attempt was made to restore the reactor performance by increasing the HRT, changing the SLR, decreasing the phenol fraction in feed and decreasing the OLR. But the sludge had to be again acclimatized to phenol. Fang et al. (2004), reported that it took 6.7 months for recovery of an UASB reactor treating phenol at ambient temperature after providing phenol shock loading for 5 days. The reactor flora was highly disturbed with the shock loading. A comparison of the present investigation with other studies is presented in Table 4.2. It can be observed from this table that this anaerobic fixed bed reactor was capable of degrading highly concentrated phenolic wastewater (3.76 g/l) successfully at stable conditions. However the reactor became unstable after increasing the phenol concentration from 3.76 to 4.7 g/l in one step. The phenol and COD efficiency dropped constantly for 1 week and sharply after that. Nonetheless the reactor was capable of removing both, COD and phenol at > 85 % efficiency for one week after this phenol increment.

After shock load, on day 295, the dissolved COD in the effluent was even more than in the influent. The microbes protect themselves from cell injury caused by toxicant, via a survival mechanism where toxicant is flushed out of the cell. This mechanism is called efflux (Kieboom 1998). In addition some of them may have lysed due to the phenol shock. Efflux is a conceivable method to effectively reduce the actual concentration of a toxic solvent for example in bacterial membranes. Efflux transporters are located in the cytoplasmic membrane and thus, in bacteria, the agents may be assumed to be pumped out into the periplasm (Nikaido, 1994). The outer membrane of gram negative bacteria has narrow channels made up of protein molecules. The efflux transporter molecule connects with the outer membrane channel protein via accessory proteins present in the periplasm. The phenol molecules can be expected to enter the bacterial cell through the porin channels resulting in their uptake. Once phenol starts accumulating inside the cell, the efflux protein flushes them out into the periplasm and then to the extra-cellular medium with the help of the accessory and porin proteins (Sharma et al., 2002). Such an efflux system was indeed identified in *Pseudomonas putida* S12 showing efflux of toluene (Isken and Debont, 1996). The accumulation of COD made it necessary to provide no feed to the reactor. However the flora was not active. Therefore after continuation of feed with phenol loading, the removal efficiency was not encouraged. The low phenol concentration of 0.19 g/l, preceded by only glucose in the feed, provided some help for revival of COD removal. After that the phenol loading was again increased gradually upto 1.32 g/l in the influent feed. When the glucose was completely omitted from the synthetic wastewater on day 450, there was less COD removal as expected from the phenol removal in the reactor. The anaerobic degradation of aromatic compounds is generally a two-phase process. In the first phase, the aromatic substrates are converted to one of a few central intermediates, which are suitable for reductive dearomatising (peripheral metabolism). The second phase of aromatic degradation is reduction of these intermediates to non-aromatic compounds which are further degraded to acetyl-CoA and CO_2 (Heider and Fuchs, 1997). Investigations carried out by Knoll and Winter (1989) and Gallert and Winter (1992, 1994) indicated that phenol was degraded via carboxylation to benzoate, 4-hydroxybenzoate, 4-hydroxybenzoyal-CoA and benzoyl-CoA by a methanogenic bacterial consortium. An obligate syntrophy between the phenol-carboxylating, benzoate-degrading and methane-producing bacteria was required for the complete phenol degradation to biogas. The decreasing efficiency of phenol and COD removal after shock loading indicated that there was a disturbance in syntrophy, leading to incomplete mineralization of phenol. However this disturbance was not permanent. COD and phenol removal could be restored within 50 days, indicating that no phenol intermediate production occurred that prevented restoration permanently.

Table 4.2: Studies of anaerobic degradation of pure and mixed phenolic compounds

Substrate	Phenol Conc. (mg/l)	Phenol COD (mg/l)	Total OLR (g COD/l'd)	Total COD removal (%)	System	Remarks	Reference
Phenol	630	1500	0.9	>99[a]	UASB[d]	Reactor operation at 55^{0}C	(Fang et al., 2006)
Phenol + Sugarcane molasses (SM)	1176	2800	8	87-99.9	UASB[d]	Reactor failure at 4:1 COD ratio (Phenol: SM)	(Gali et al., 2006)
Phenol	1260	-	6	98[a]	UASB[d]	Reactor operation at 26^{0}C. OLR > 6 caused deterioration in removal efficiency (26 ^{0}C operating temperature)	(Fang et al., 2004)
Phenol + p-Cresol	800-1200	1000-4000[b]	7	> 90 %	UASB[d]	Sharp decrease in removal efficiency at influent phenolic[b] conc. of 5g COD/l	(Flores et al., 2003)
Phenol	1260	-	6	86	UASB[d]	-	(Tay et al., 2001 b)
Phenol	1200	-	3	99[a]	HAIB[d]	Reactor operated for only 1 week at phenol conc. 1200 mg/l	(Bolanos et al., 2001)
Phenol + p-Cresol	-	-	0.66	85	Bioaugmented enriched consortium	-	(Tawfiki et al., 2000)
Phenol + m-Cresol + o-Cresol	-	-	2.12	94	Fixed film UAB[d]	-	(Tawfiki et al., 1999)
Phenol + m-Cresol	-	-	4.3	98[a]	UASB[d]	-	(Zhou and Fang, 1997)
Phenol + Formaldehyde + Methanol	-	-	3.4	95	Anaerobic activated carbon	-	(Goeddertz et al., 1990)
Phenol + Glucose	3764	9000[c]	5.3	90 and 93.7[a]	AFBR[d]	Steady state removal	Present work
Phenol + Glucose	4705	11200[c]	5.7	91 and 89.5[a]	AFBR[d]	Reactor was not stable and sharp deterioration in removal efficiency was followed within a week	Present work

Where, a- phenol COD ; b- phenol + p-Cresol ; c- 1 g phenol = 2.38 g COD; d- UASB-Upflow anaerobic sludge blanket reactor, HAIB-Horizontal flow anaerobic immobilized biomass reactor, UAB- Upflow anaerobic reactor and AFBR- Anaerobic fixed bed reactor.

4.3 Phenol biodegradation in batch reactors

4.3.1 Aerobic batch assays

4.3.1.1 Phenol degradation profile

The phenol removal rate increased considerably in the shake flasks with consecutive feedings. Many researchers compared phenol removal using batch cultures of free cells and attached growth systems in laboratory experiments. For instance, a three membered aerobic phenol-degrading consortium, containing *Pseudomonas fluorescens* III strain, *Acinetobacter* and a spiral shaped strain had degraded 376 mg/l of phenol during 20 h (Khoury et al., 1992). Knoll and Winter (1989) reported a four member anaerobic consortium capable of degrading 188 mg/l in 6 h, whereas phenol degradation by a mixed consortium in a fed batch reactor was reported to degrade 480 mg/l during 12 h (Ambujom and Manilal, 1995). Kumar et al. (2005) reported 1000 mg/l phenol removal in 162 h by phenol/catechol acclimatized *Pseudomonas putida* MTCC 1194 in shake flasks. Esparza et al. (2006) observed 19 to 51 % removal of 2 g/l phenol in 6 h with different influent medium but the same phenol concentration in the presence of phenol acclimatized biomass. Their study indicated that low medium conditions could affect removal rates. However their highest phenol removal rate was comparable to batches that were carried out in the present study (Section 3.2.1.1 and 3.2.1.2.1). Whereas Marrot et al. (2006) reported that a period of 54 h was required for complete degradation of 3 g/l phenol in batch assays. Gouder et al. (2000) observed no phenol degradation for concentrations above 1.3 g/l with mixed cultures adapted to phenol. Therefore, results from the present work were very promising and indicated that high biological phenol degradation rates in suspended cultures (8.84 g/l·d to 4.85 g/l·d) could be achievable with a previously adapted culture obtained from an immobilised system.

4.3.1.2 Acclimatised sludge and enrichment culture

Another set of batch assays was performed to study the behaviour of phenol degradation in the presence of co-substrates with different inocula (adapted sludge, enrichment culture and non-adapted sludge). In the phenol adapted sludge, glucose, acetate and 2-CP were added as co-substrates along with phenol at the start time i.e. t=0. From the rapid degradation of glucose and acetate, it was clearly evident that the phenol adapted sludge contained microorganisms capable of degrading glucose and acetate. The preferred C- sources were glucose and acetate, since the phenol degradation rate was reduced in both cases as compare to the assay with only phenol as carbon source.

In the enrichment culture (obtained from the 6-7 transfers of the adapted sludge into a phenol containing mineral medium), the effect of co-substrate addition was investigated once the cells were actively metabolising phenol after overcoming the initial lag phase. A marked reduction in the phenol utilisation rate in the presence of acetate was observed. Glucose as a co-substrate was scarcely metabolised but it did not hinder the phenol utilisation rate. The lack of glucose utilisation in the enrichment cultures may be due to the fact that phenol degrading microorganisms were not able to metabolise glucose and glucose metabolising bacteria were lost in the specifically enriched consortium. There was a difference in glucose utilisation between the adapted sludge and the specifically enriched cultures. It was evident that glucose addition had an effect on phenol removal in phenol adapted sludge (fed on phenol + glucose), but had no effect on phenol removal in the enrichment cultures fed only phenol. This finding with glucose as a co-substrate was similarly reported by Kar et al. (1996). They added glucose and phenol individually and in mixtures in 3 batch reactors, respectively, containing a phenol acclimatizised culture of *Arthrobacter sp*. It was noticed that growth of biomass accompanied by phenol degradation started immediately in reactor 2 and 3 which contained phenol as one of the substrates and no growth was observed in reactor 1 for a long time (8-10 h) as no significant glucose removal occurred for 10 h. No previous studies are available reporting acetate utilisation after phenol degradation in phenol acclimatised enrichment cultures. The addition of acetate, both after or before the end of the lag phase for phenol degradation reduced the biodegradation rates for phenol. In the literature, there are some studies on the effect of co-substrates on phenolic compounds degradation reporting the enhancement of toxic substance degradation (Loh and Wang, 1998; Yu and Ward, 1994). Lob and Tar (2000) reported that the rate of phenol removal was increased when 1 g/l glucose was supplemented in addition due to better growth of additional cell mass which enhanced the phenol removal. However the improvement in phenol removal was only 13 %. When the concentrations of glucose were exceeding 1 g/l, the degradation rate of phenol dropped below the rate achieved in the absence of glucose. In the present study with phenol adapted sludges, the later behaviour was observed. It may be attributed to the catabolic repression of the phenol degradation pathway by co-substrates in order to enhance biomass acclimatisation to alternate carbon sources. This phenomena has been well documented by other researchers (Chitra et al., 1995; Satsangee and Ghosh, 1990; Rozich and Colvin, 1985). Different easily degradable co-substrates like glucose and acetate could produce different effects on phenolic compound degradation. Fakhruddin and Quilty (2005) had investigated 2-CP removal in the presence of glucose and fructose. The degradation was simulated in the presence of low concentrations of glucose (0.05-1.0 % w/v) and removal was inhibited due to significant decrease of the pH with concentrations of glucose greater than 1.0 % (w/v). This was alleviated when the pH was controlled; however the rate of removal was greater in the presence of fructose than in the presence of glucose.

In the present study, the addition of 2-CP to phenol adapted sludge retarded the phenol degradation rate. This may be due to the fact that 2-CP cannot support good cell growth and that it is inhibitory for the oxidation of phenol. Inspite of this, the sludge did not loose its phenol metabolising capacity even at toxic concentrations of 2-CP. After slow phenol biotransformation and degradation, the sludge began to remove 2-CP as well, indicating that phenol acclimatised sludges can degrade other toxic phenolic compounds (Fig. 3.31 d). Loh and Wang (1998), investigated biodegradation of phenol in the presence of 4-chlororphenol (4-CP) at various concentrations and reported that 4-CP, not only severely inhibited cell growth and reduced the yield of biomass but also adversely slowed down degradation of phenol.

4.3.1.3 Unacclimatised sludge

The aerobic biodegradation of the phenol proceeds via two pathways (Fig. 1.2). Before complete mineralization, a number of phenol intermediate metabolites are produced. The COD and SAC_{254} data (Table 3.7) after 24 hours of incubation of fresh activated domestic sewage sludge with phenol indicated, that, except for assay no. IV, metabolites from incomplete degradation of phenol were present in assay I, II and III. Acetate and glucose were completely removed within 24 h. Thus after 24 h there was no additional carbon source present in the assays other than phenol. This was confirmed by measuring the SAC_{254}. The 2-CP addition had a negative effect on the sludge. There was no change either in phenol or in 2-CP concentration after 1 day and even after 1 week of incubation (data not shown in Table 3.7). Dionisi et al. (2006), carried out test for the biodegradation of various organic xenobiotics in unacclimatized activated sludge and found that only phenol was biodegraded in 5 h with initial concentration of 17 mg/l, showing that the potential of unacclimatised activated sludges to biodegrade aromatic xenobiotics was rather limited. In the present study, phenol was not found after 24 h, may be due to its conversion to metabolic intermediates in the presence as well as in the absence of glucose and acetate. Adsorption to cells was ruled out as evident from the remaining dissolved total COD. In assay II and III, the COD removal was greater than in assay I because of degradation of the co-substrates acetate and glucose (Table. 3.7). This is a very interesting phenomenon of conversion of 0.19 g/l phenol into intermediates in presence of co-substrates with activated sludge in 24 h. It has not been reported elsewhere.

4.3.1.4 Biodegradation kinetics

Haldane kinetic constants were calculated in an aerobic batch assay. The value of K_i obtained in the present study was comparatively higher for mixed culture (Table 4.3). A higher value of K_i (up to 2434.7 mg/l) than in this study was calculated by Artuchelvan et al. (2006) for a culture of *Bacillus brevis,* which was very unusual for a pure culture. No other report indicating such high values with a pure culture was found. The value of μ_{max} obtained from data of the present investigation was comparable to other studies with mixed cultures, however the value of K_s was little higher, indicating that microorganisms were able to grow at relatively high concentrations (Table 4.3). The Y obtained at different concentrations (Table 3.9) is comparable at all concentrations except for the lowest concentration of 0.25 mmol/l due to the small amount of substrate. The phenol depletion was very fast at low concentrations but still the rate of degradation per hour, calculated from the experimental data was less at low substrate concentrations due to the lower amount of initial substrate which was utilised very quickly.

Table 4.3: Haldane kinetic constants cited in literature and from present study

Microbes	Max. phenol conc.(mg/ l)	μ_{max} (h)	K_s (mg/ l)	K_i (mg/ l)	Temp (^{0}C)/ pH	Reference
Bacillus brevis	750-1750	0.026-0.078	2.2-29.3	868-2434.7	34±1/-	Arutchelvan et al., 2006
Candida tropicalis		0.48	11.7	207.9	-	Yan et al., 2005
Pseudomonas putida MTCC 1194	1000	0.305	36.33	129.79	29.9±0.3/-	Kumar et al., 2005
Trichosporon cutaneum R57	-	0.42	110	380	-	Alexievaa, 2004
Pseudomonas putida F1 ATCC 700007	50	0.051	18	430.0	30/7.0	Abuhamed et al., 2004
Acinetobacter	350	0.83	1.5	250	30/-	Hao et al., 2002
Pseudomonas putida DSM 548	100	0.436	6.19	54.1	26/6.8	Monteiro et al., 2000
Acinetobacter calcoaceticus	0-500	0.542	36.2	145.0	-	Kumaran and Paruchuri, 1997
Pseudomonas fluoroescens 2218	0-600	0.618	71.4	241.0	-	Kumaran and Paruchuri, 1997
Pseudomonas. putida P 71	-	0.569	18.5	99.4	25/6.8	Beyenal et al., 1997
Pooled culture (*P.fluorescens, P. putida, P. cepacia, A. calcoaceticus, C. tropicalis*)	0-500	0.452	53.9	516.0	-	Kumaran and Paruchuri, 1997
NCIB8250 (*Acineto bacter sp.*)+ NCIB10535 (*Pseudomonas sp.*)+ NCIB1015 (*Pseudomonas sp.*)	0-500	0.418	2.9	370.0	30/6.7-6.9	Livingston and chase, 1989
Mixed culture	0-800	0.308	44.92	525.0	27±1 /7.0	Saravanan et al., 2008
Mixed culture	2500	0.438	29.5	72.4	Ambient/ 6.5	Marrot et al., 2006
Mixed culture	-	0.131	5.0	142.0	-	Adamo et al., 1984
Mixed culture	40	0.258	3.9	121.7	15/-	Buitron et al., 1998
Mixed culture	0-900	0.260	25.4	173.0	28±0.5/6.6	Pawlowsky and Howell, 1973
Mixed culture	23.5-659	0.3095	74.65	648.13	25± 2/7.2	This study*

4.3.2 Anaerobic batch assays

The metabolic pathway of anaerobic phenol degradation (Fig. 1.3) involves a number of aromatic and non-aromatic intermediates, which tend to accumulate in case of incomplete mineralization. The aromatic ring cleavage is considered to be the rate limiting step in the phenol degradation. The complete cleavage of the ring structure ensures completion towards mineralization process of the byproducts to the end products like acetate or/and CH_4 and CO_2. As indicated earlier, it was difficult to determine the intermediates with the available gas chromatographic method; therefore measurement of SAC_{254} was taken as an indication of ring cleavage. The glucose fed inoculum taken from the suspended bed reactor was able to mineralized benzoic acid (intermediate of phenol), when given as substrate, as well as phenol without any accumulation of aromatic intermediates. It has been reported previously that phenol degradation proceeds via benzoate (Knoll and Winter, 1987). However no significant intermediate accumulation during batch phenol degradation up to 376 mg/l concentrations with an enriched consortium was also observed by Khoury et al. (1992). The presence of phenol up to 0.19 g/l concentration did not inhibit methanogenesis in the test inoculum taken from the suspended bed reactor indicating the ability of the reactor flora to degrade phenol at degradation rates of about 295 mg/l·d phenol.

The effluent from the phenol degrading anaerobic fixed bed reactor used as inoculum in an another suspended batch assay was able to degrade phenol with much higher rates of 966 mg/l·d indicating the high adaptation ability of the microbial flora under immobilised conditions. In a phenol degradation batch test conducted by Fang et al. (2005), it took > 15 days for complete phenol removal at a 400 mg/l concentration. The seed inoculum was the sludge taken from the UASB reactor degrading 100 % of 600 mg/l phenol at thermophilic conditions; this may have been due to the selection of different microbial communities and/or slow activity of the enzymes involved in degradation of phenol at thermophilic temperatures (Leven et al., 2005). However the batch test carried out at ambient temperature by the same group (Fang et al., 2004) with the sludge taken from the UASB treating 1.26 g/l phenol at 12 h HRT also took ca. 4 days to remove 400 mg/l phenol.

4.3.3 Anoxic batch assays

Under reducing conditions, the anoxic batch assay demonstrated a much higher removal rate (1.2 g/l·d) compared to the anaerobic batch assay carried out with the inoculum taken from the anaerobic suspended bed reactor. It has been reported that the phenol degradation rate of denitrifiers is higher than that of methanogenic consortia (Fang et al., 1999). Also anoxic sludge was able to tolerate much higher phenol concentrations under batch conditions. This was due to better adaptation to phenol in the anoxic suspended bed reactor than in the anaerobic suspended bed reactor. The simultaneous removal of nitrate with phenol (Fig 3.41) indicated a successful denitrification process, where phenol acted as electron donor. However all nitrate was not removed due to lack of enough carbon source as explained previously (Section 3.23). At 2 mmol/l (0.19 g/l) phenol concentration, denitrifying 1 g NO_3^--N required 3.8 g COD which was near the average ratio of 3.5 observed in the continuous reactor studies (Section 3.1.1.3.3).

4.4 Chlorophenol degradation

4.4.1 Acclimatisation of sludge to 2-CP in a batch reactor

Due to its chemical structure, chlorophenols are more difficult for biodegradation than phenols. Therefore it was important to check the degradation of 2-CP in the sewage sludge and thus acclimatise it in order to get a successful culture for the start-up of the reactor. Only 23 days were required to acclimatise the sewage sludge to 2-CP. Wang et al. (1998) reported 35 days as the time required for the complete transformation of only 10 mg/l of 2-CP in a methanogenic enrichment culture of a wastewater treatment plant. Quick successive feeding of 2-CP upon complete substrate depletion reduced significantly the lag time of each successive degradation phase. Ye and Shen (2004) performed chlorophenol acclimatisation studies on 2 types of sludges under batch conditions. They had also observed a relatively rapid degradation rate of 2-CP with municipal wastewater sludge after 2 months of incubation with successive feedings.

4.4.2 Performance of anaerobic fixed bed reactor (AFBR) for 2-CP degradation

The removal efficiency at the reactor start-up was high (Fig 3.42) because of the prior acclimatisation of the bacterial sludge to the use of 2-CP as the sole carbon source. Only a minimal amount of phenol (an intermediate product of reductive dechlorination) was detected in the effluent during the continuous reactor operation. The phenol was considered to be readily degraded as soon as it was derived from the 2-CP dechlorination, indicating the presence of phenol degrading bacteria in the reactor sludge. Thus, the bioreactor exhibited a favourable ability to dechlorinate 2-CP. Consistent to this study accumulation of any dechlorination intermediates was also not reported by Majumder and Gupta (2007), Chang et al. (2004) and Bae et al. (2002). However during batch acclimatization of sewage sludge to 2-CP phenol, accumulation was observed for a short time period (Fig. 3.41). This may be due to the rapid rate of metabolite utilisation in continuous reactors compared to batch reactors during the acclimatization phase. Absence of phenol during reductive dechlorination of 2-CP in the continuous mode and it's presence during batch mode was also reported by Chang et al. (2003). However the mechanism of 2-CP degradation in the present study could differ from the study of Dietrich and Winter (1990), who observed high levels of phenol in the effluent during 2-CP degradation in a continuous fixed bed reactor.

The COD removal (Fig. 3.45) was clearly due to the 2-CP removal (Fig 3.43) indicating that the microbial flora which was well adapted to 2-CP had a preference to 2-CP. The effect on reactor performance due to changes in feed was minimal. This indicated that the organic nitrogen sources could be replaced by inorganic nitrogen without any kind of negative effect on the microbial flora or reactor operation. Kafkewitz et al. (1996) observed that in the presence of a vitamin solution, biomass growth was increased during the dechlorination of 2-CP. A similar result was reported by Smith and Woods (1994) who found that the presence of Vitamin B_{12} favoured the reductive dechlorination of various chlorophenols. Successful treatment of 2-chlorophenol at such high concentration has not been reported before (Table 4.4).

The average ratio of g chloride released per g 2-CP dehalogenation was 0.20. In a study by Majumder and Gupta (2007) the observed chloride released during the biodegradation of 2-CP in an UASB reactor was in the range of 0.22-0.24. This is also in confirmation with Zilouei et al. (2006), who carried out experiments on biodegradation of a mixture of chlorophenols (2-CP; 4-CP; 2, 4-DCP and 2, 4, 6-TCP) in packed-bed bioreactors using mixed bacterial consortia under aerobic conditions and different loading rates. Under all conditions, chloride was released to 84 $\pm$ 9 % of the theoretical value calculated from the amount of degraded chlorophenols.

4.4.3 Batch Assay

The highest 2-CP degradation rate (175.2 mg/l·d) obtained in batch assays (Fig 3.49) was very close to 180 mg/l·d observed in suspended cultures by Dietrich and Winter (1990). However, in the present study the highest degradation rate observed in the batch experiment was almost 5 times less than what was observed in the continuous reactor. The reason was the inoculum concentration, as in the AFBR a thick layer of sludge around the carrier material was clearly visible, whereas the effluent from the reactor lacks the sludge and contained only the few microrganism that were present in clear suspension, indicating far less amount of inoculum in the batch assays compared to the population density of the reactor. In another assay the presence of glucose inhibited the 2-CP mineralization process because the sludge fed with glucose-free but 2-CP containing mineral medium was selective for the 2-CP degrading bacteria over a long time period in the reactor.

Table 4.4: Performance of various systems for 2-CP degradation

Reactor	2-CP Concentration (mg/l)	2-CP- Loading rate (2-CP mg/l·d)	2-CP - Removal rate (mg/l·d)	Remarks	Reference
Suspended batch	-	-	180	Enrichment culture	Dietrich and Winter (1990)
AFBR[a]	2000	70-600	375	Phenol accumulation observed	Dietrich and Winter (1990)
SMBR[b]	25	40	37	Hydrogenotrophic conditions	Chang et al. (2003)
Batch	12.8	-	-	Denitrifying conditions	Bae et al. (2002)
Batch	10-25	-	about 0.3 (maximum)	Experiment carried out in presence and absence of sucrose	Ye and Shen (2004)
UASB[c]	30	45-120	117	3 g/l sodium acetate used as co-substrate	Majumder and Gupta (2007)
Suspended batch	50-192	-	115-175	Experiment carried out with acclimatised inoculum	Bajaj et al. (2008)[1] Present study
AFBR	250-2600	50-1365	873 (maximum)	The stable removal rate achievable was 730 mg/l·d	Bajaj et al. (2008)[1] Present Study

a-AFBR- Anaerobic fixed bed reactor
b-SMBR- Silicone membrane bioreactor
c-UASB-Upflow anaerobic sludge blanket

Chapter 5

Summary

Studies were carried out with suspended and fixed bed reactors to investigate biological phenol degradation. These investigations pointed out the better efficiency of fixed bed reactors over suspended bed reactors for microbial phenol removal. Therefore the further study on microbial 2-chlorophenol removal was carried out only in a fixed bed reactor. The highest phenol and 2-chlorophenol removal rates in the continuous reactors are presented in Table 5.1

Table 5.1: Performance of investigated continuous reactors

Reactor type	Max. removal rate (g/l·d)	Substrate		OLR (COD g/l·d)		HRT (d)
		Type	Conc. (g/l)	Total	From phenol/2CP	
Biofilm: Aerobic	2.93	Phenol	4.90	15.30	12.92	0.95
Biofilm: Anaerobic	1.62	Phenol	4.70	5.68	4.47	2.50
Suspension: Anoxic	0.75	Phenol	0.75	5.90	2.0	0.90
Suspension: Aerobic	0.55	Phenol	0.75	4.13	1.54	1.16
Suspension: Anaerobic	0.44	Phenol	1.13	2.18	1.15	2.34
Biofilm: Anaerobic	0.87	2-CP	2.60	2.33	1.91	2.20

5.1 Suspended bed reactors

Glucose was used as the main carbon substrate for the start-up of the reactors to initiate biomass growth and to simulate industrial wastewater with 5 g/l COD concentration. The results obtained in this study seem to suggest that anaerobic and aerobic sludges grown on glucose for long time periods in the suspended bed reactors were sensitive to the presence of an aromatic organic compound such as phenol. These reactors were prone to quick unsteady state as a response to change in conditions such as accidental oxygen deficiency (aerobic reactor) or thermostat breakdown resulting in unregulated low temperatures (anaerobic reactor), and required long stabilisation times. In particular, reactors were sensitive to increases of the phenol concentration. Starvation of anaerobic sludge had helped in recovery by utilisation of accumulated fatty acids.

Performance of anoxic suspended bed reactors in terms of high organic loading was much better than that of aerobic or anaerobic suspended-bed reactors. This reactor had an OLR of average 6 g COD/l·d and could sustain organic loads up to 10 g COD /l·d; however presence of high concentration of phenol in influent was not favourable. The COD and nitrate removal efficiencies were independent of increases in OLR but dependent on phenol concentrations. The decrease in COD removal was reversible as soon as the influent phenol concentrations were lowered to previous conditions. The reactor demonstrated a successful denitrification process with an average COD : NO_3-N ratio of 3.5. A phenol degradation rate of 1.2 g/l·d at a concentration of 0.19 g/l was achievable in batch studies carried out with the sludge obtained from the continuous reactor.

5.2 Fixed bed reactors

Among all reactors, the highest phenol removal rate was achieved in an aerobic fixed bed reactor at a phenol/COD ratio of 0.8 and an influent phenol concentration of 4.9 g/l, obtained after stepwise small increments of the influent phenol concentration. Thus, it is feasible to use aerobic biological treatment of wastewater with phenol concentrations as high as 5 g/l. The stability and efficiency of this process depended upon OLR. Phenol removal showed a nearly identical trend to that observed for the total COD in the wastewater, indicating that no intermediate was enriched. The phenol acclimatised aerobic flora responded negatively in terms of phenol degradation in the absence of nitrogenous substances, even for a short period of time. The phenol degradation rates increased again as soon as nitrogen supply was resumed. In shake flask batch assays of phenol with acclimatised sludge, the phenol removal rate improved considerably with consecutive feeding, provided that the batch was operated under the conditions in which the degrading capacity of the biomass was limited only by the availability of phenol and not by the availability of other nutrients. The highest phenol removal rate observed in batch assay was 1.51 g/l·d in the presence of acetate and it increased to 3.54 g/l·d (more than in the continuous reactor) after acetate depletion. Additionally, the effect of co-substrates on removal of phenol was investigated using acclimatised and unacclimatised aerobic biomass. With acclimatised biomass taken from the reactor and incubated in shake flasks in mineral media, the easily degradable co-substrates glucose and acetic acid reduced the phenol biodegradation rate. The not easily degradable co-substrate 2-CP attenuated the phenol degradation rate but the sludge was capable of also removing 2-CP, once the phenol was degraded. An enrichment culture that was obtained after several transfers into a phenol containing mineral medium degraded preferentially phenol, even in the presence of glucose which did not affect the phenol removal rate. When phenol and acetic acid were supplied together, the enrichment culture showed concomitant phenol and acetic acid removal. However the phenol degradation rate was attenuated, independent of the time of its addition into the assay- before or after the lag phase of phenol degradation. The different behaviour of a more complex adapted sludge and specifically on phenol cultured enrichment was not reported before. The use of enrichment cultures from the adapted sludge provides an advantage by ensuring uninterrupted phenol degradation in the presence of glucose.

Also very interesting results were obtained with unacclimatised sludges where no phenol was detected in assays after 24 h. The COD and high SAC_{254} values in assays indicated however, that phenol was only converted to some unknown aromatic metabolites. The data of a batch assay carried out with the sludge of the aerobic fixed bed reactor fitted very well with Haldane kinetic model, giving a non linear regression with a correlation coefficient of 0.987. The constants derived from the Haldane equation were $\mu_{max} = 0.3095$ h, $K_s = 0.7933$ mmol/l (74.65 mg/l) and $K_i = 6.887$ mmol/l (648.13 mg/l). The inhibition constant for phenol was higher than in most previous studies with mixed cultures (Table 4.3), indicating a better adaptation or a higher tolerance to phenol of the mixed culture taken from the aerobic fixed bed reactor.

The anaerobic mixed microbial flora could be adapted to degradation of high phenol concentrations of 3.76 g/l in about 200 days with an average of 94 % phenol removal in an anaerobic fixed bed reactor. The steady state phenol removal observed before the phenol shock load was 1.48 g phenol/l·d with an influent phenol concentration of 3.7 g phenol/l at a HRT of 2.5 days and an OLR of 5.3 g/l·d. The COD removal before exposing the reactor to the shock load was 93.4 % indicating the complete degradation of phenol as it corresponded with phenol removal. The reactor was stable at high phenol concentrations only when the influent phenol concentration was increased in 0.19-0.28 g/l increments. An increment of 0.94 g/l of phenol was fatal for the stability of reactor. After the shock load a consistent drop in COD and phenol removal was observed for the first week, followed by a sharp decline and production of fatty acids. Recovery of the reactor was possible only when no feed was provided to the reactor for one month and the phenol concentration was kept as 0.19 g/l at restart. The phenol concentration was then gradually increased to 1.32 g/l at different OLR's. At stable operating conditions of a reactor without glucose in the feed (that was previously fed glucose and phenol), intermediates of anaerobic phenol metabolism were observed for a period of 50 days after which total dissolved COD and phenol removal in the effluent were in balance.

Very high 2-CP degradation rates in anaerobic enrichments of sewage sludge have been obtained in this study. The degradation of 50 mg/l of 2-CP was possible in ca. 3 weeks. The removal efficiency increased considerably with successive feedings. The acclimatised sludge could be used as an inoculum for an anaerobic fixed-bed reactor to achieve high-rate 2-CP degradation under continuous fermentation conditions. A stepwise increase of 2-CP in the medium finally allowed successful continuous degradation of 2-CP up to 2.6 g/l. This was the highest 2-CP influent concentration ever tested. The highest permanently achievable, "safe" degradation rate of 2-CP in this reactor was around 0.73 g/l·d. A negative effect of 2-CP shock loading at 1.36 g/l·d was reversible after 2 days of starvation and a set back of the 2-CP-loading. Within 2 weeks the 2-CP removal efficiency recovered and reached again 80 %. This indicated that inhibitory effects due to high chlorophenol loading were recoverable if such loading was short term and previous loading conditions were followed. Phenol, an intermediate of 2-CP biodegradation, was observed during the acclimation of sewage sludge under batch conditions after each feeding, but no phenol was detected in the effluent during continuous reactor operation, indicating a quick utilisation of the aromatic intermediate.

Organic nitrogen and vitamin sources such as peptone and yeast extract could be replaced by NH_4NO_3 and a defined vitamin solution without a permanent negative effect on the microbial flora or reactor operation. The average ratio of chloride released per unit of 2-CP removed was 0.20 g Cl⁻ per g 2-CP added. The median value was 0.24, which was close to the theoretical value of 0.28 g Cl⁻ per g 2-CP. This indicated the successful dechlorination of 2-CP. A 2-CP degradation rate of 175 mg/l·d was achievable within a batch assay containing effluent of the reactor as the inoculum. In the batch studies, the presence of yeast extract or peptone did not affect the degradation but when glucose was provided as co-substrate, rates were considerably decreased.

It is concluded from this study that the acclimatisation strategy seems to play an important role in successful biological treatment of toxic compounds such as phenol and 2-CP at high concentrations. Once a suitable microbial population was established for aerobic/anaerobic, phenol or 2-CP degradation in the biofilm of fixed bed reactors, deleterious conditions for suspended bed reactors, such as sludge washouts, overloading, temperature breakdown, clogging etc. may be overcome in the fixed beds without a complete breakdown of the process.

Zusammenfassung

Um den biologischen Phenolabbau zu untersuchen wurden Studien mit Schwebebettreaktoren und Festbettreaktoren durchgeführt. Hierbei zeigte sich die größere Effizienz der Festbettreaktoren gegenüber den Schwebebettreaktoren. Die weiteren Versuche zum mikrobiellen 2-Chlorphenolabbau wurden daher in einem Festbettreaktor durchgeführt. Die höchsten Phenol und 2-Chlorphenol Abbauraten in den Durchflussreaktoren sind in Tabelle 5.1 dargestellt.

Tabelle 5.1: Betriebsverhalten der untersuchten Durchflussreaktoren

Reaktor typ	Max. Abnahmerate (g/l·d)	Substrat		OLR (COD g/l·d)		HRT (d)
		Typ	Konz. (g/l)	Gesamt	Von phenol/2CP	
Biofilm: Aerob	2.93	Phenol	4.90	15.30	12.92	0.95
Biofilm: Anaerob	1.62	Phenol	4.70	5.68	4.47	2.50
Suspension: Anoxisch	0.75	Phenol	0.75	5.90	2.0	0.90
Suspension: Aerob	0.55	Phenol	0.75	4.13	1.54	1.16
Suspension: Anaerob	0.44	Phenol	1.13	2.18	1.15	2.34
Biofilm: Anaerob	0.87	2-CP	2.60	2.33	1.91	2.20

5.1 Schwebebettreaktoren

Bei der Inbetriebnahme des Reaktors wurde Glukose als Haupt-Kohlenstoffsubstrat verwendet um das Wachstum der Biomasse einzuleiten und um industrielles Abwasser mit einer COD Konzentration von 5 g/l zu simulieren. Die im Rahmen dieser Untersuchung erhaltenen Ergebnisse weisen darauf hin, dass anaerobe und aerobe Schlämme, die über längere Zeiträume auf Glukose in den Schwebebettreaktoren angezogen wurden, sensitiv auf die Anwesenheit von aromatischen, organischen Substanzen wie zum Beispiel Phenol reagierten. Diese Reaktoren zeigten sich anfällig für einen schnellen Verlust des Fließgleichgewichts aufgrund einer Änderung der Bedingungen wie z.B. versehentlichem Sauerstoffmangel (aerober Reaktor) oder einem Ausfall des Thermostats, der zu unregulierten, niedrigen Temperaturen führte (anaerober Reaktor). Ferner benötigten sie lange Stabilisationszeiten. Im Besonderen reagierten Reaktoren sensitiv auf eine ansteigende

Phenolkonzentration. Das Aushungern des anaeroben Schlamms half bei der Erholung durch Nutzung von akkumulierten Fettsäuren.

Bei Betrieb der anoxischen Schwebebettreaktoren erzielte man bei hoher organischer Belastung deutlich bessere Ergebnisse als bei den aeroben oder anaeroben Schwebebettreaktoren. Dieser Reaktor wies im Normalfall eine OLR von 6 g COD/l·d auf und konnte organische Belastungen von bis zu 10 COD g/l·d ertragen. Dennoch war eine hohe Phenolkonzentration im Zufluss nicht vorzuziehen. Die Effizienz der COD und Nitrat Entfernung war unabhängig von einem Anstieg der OLR aber abhängig von der Phenolkonzentration. Das Sinken der COD-Abbaurate bei der Steigerung der Phenolkonzentration war reversibel sobald die Phenolkonzentration im Zufluss wieder auf die vorherigen Bedingungen reduziert wurde. Der Reaktor zeigte einen erfolgreichen Denitrifikationsprozess mit einem normalen COD : NO_3-N Verhältnis von 3,5. Dabei konnte eine Phenolabbaurate von 1,2 g/l·d bei einer Konzentration von 0,19 g/l in Batch-Tests mit Schlamm aus dem Durchflussreaktor erreicht werden.

5.2 Festbettreaktoren

Von allen Reaktoren wurde die höchste Phenolabbaurate in einem aeroben Festbettreaktor erreicht. Dabei lag das Phenol/COD-Verhältnis bei 0,8 und die höchste Phenolkonzentration im Zufluss betrug, nach einer Steigerung der Phenolkonzentration in kleinen Schritten, 4,9 g/l. Daher ist eine aerobe biologische Abwasserbehandlung bei Phenolkonzentrationen bis zu 5 g/l möglich. Die Stabilität und Effizienz dieses Prozesses ist abhängig von der OLR. Der Phenolabbau verlief parallel zum COD-Abbau im Abwasser. Das weist darauf hin, dass kein Zwischenprodukt angereichert wurde. Die an Phenol akklimatisierte aerobe Flora reagierte selbst über kurze Zeiträume negativ bezüglich des Phenolabbaus in Anwesenheit von stickstoffhaltigen Substanzen. Die Phenolabbaurate stieg wieder an, sobald die Stickstoffversorgung wieder aufgenommen wurde. In Schüttelkolben Batch-Tests von Phenol mit akklimatisiertem Schlamm stieg die Phenolabbaurate mit fortlaufender Fütterung merklich an. Voraussetzung dafür war, dass die Bedingungen des Batch-Tests so gewählt wurden dass die Abbaukapazität der Biomasse nur von der Phenol-Verfügbarkeit und nicht von der Verfügbarkeit der anderen Nährstoffe limitiert wurde. Die höchste Phenolabbaurate, die im Batch-Test beobachtet wurde lag bei 1,51 g/l·d in Anwesenheit von Acetat und stieg nach der Acetat-Zehrung auf 3,54 g/l·d an (mehr als im Durchflussreaktor). Zusätzlich wurde der Effekt von Co-Substraten auf den Phenolabbau untersucht, wobei akklimatisierte und unakklimatisierte aerobe Biomasse benutzt wurde. Bei der akklimatisierten Biomasse, die aus dem Reaktor entnommen und in Schüttelflaschen mit Mineral-Medium inkubiert wurde, reduzierten die einfach abbaubaren Co-Substrate Glukose und Acetat die Phenolabbaurate. Das nicht einfach abbaubare Co-Substrat 2-CP vermindert die Phenolabbaurate, aber wenn das Phenol einmal abgebaut war konnte der Schlamm ebenso 2-CP entfernen. Eine Anreicherungskultur die nach mehreren Überführungen in ein phenolhaltiges Mineralmedium gewonnen wurde baute vorwiegend Phenol ab, sogar in Anwesenheit von Glukose, welche die Phenolabbaurate nicht beeinträchtigte. Wurden Phenol und Acetat zusammen zugegeben, zeigte die Anreicherungskultur gleichzeitig einen Phenol- und Acetatabbau. Nichtsdestotrotz wurde die Phenolabbaurate abgeschwächt, unabhängig vom Zeitpunkt der Zugabe – vor oder nach der lag-Phase des Phenolabbaus. Von einem unterschiedlichen Verhalten eines komplexeren, angepassten Schlamms und einer auf Phenol kultivierten Anreicherung wurde bislang nicht berichtet. Die Nutzung von Anreicherungskulturen vom angepassten Schlamm zeigt sich von Vorteil, durch Sicherung des ununterbrochenen Phenolabbaus in Anwesenheit von Glukose.

Ebenfalls sehr interessante Ergebnisse wurden mit unakklimatisiertem Schlamm erzielt, wo nach 24 Stunden in den Proben kein Phenol gefunden wurde. Die COD und die hohen SAK_{254} Gehalte in den Proben wiesen darauf hin, dass Phenol nur in unbekannte aromatische Metaboliten umgewandelt wurde. Die Daten der Batch-Proben mit Schlamm aus dem aeroben Festbettreaktor passten sehr gut ins Haldane-Kinetik-Modell, mit einer nicht linearen Regression mit einem Korrelationskoeffizienten von 0,987. Die Konstanten die aus der Haldane-Gleichung ermittelt wurden waren μ_{max} = 0,3095 h, K_S = 0,7933 mmol/l (74,65 mg/l) und K_i = 6,877 mmol/l (648,13 mg/l). Die Inhibitionskonstante war höher als in den meisten vorherigen Versuchen mit gemischten Kulturen (Tabelle 4.3), was auf eine bessere Adaption oder eine höhere Phenoltoleranz der gemischten Kultur aus dem Festbettreaktor hinweist.

Die anaerob gemischte mikrobielle Flora kann zum Abbau hoher Phenolkonzentrationen von 3,76 g/l in etwa 200 Tagen, mit einem durchschnittlich 94%igen Phenolabbau, in einem anaeroben Festbettreaktor geeignet sein. Das Gleichgewicht des Phenolabbaus, das vor der Phenol Schock-Belastung beobachtet wurde, lag bei 1,48 g Phenol/l·d bei einem Phenol-Zufluss von 3,7 g Phenol/l und bei einer HRT von 2,5 Tagen und einer OLR von 5,3 g/l·d. Der COD Abbau lag, bevor der Reaktor der Schock-Belastung ausgesetzt wurde, bei 93,4%. Das deutete auf vollständigen Phenolabbau hin, da der COD und der SAK mit parallel abnahmen. Der Reaktor war bei hohen Phenolkonzetrationen nur stabil wenn die zufließende Phenolkonzentration in 0,19 – 0,28 g/l Schritten erhöht wurde.

Ein Schritt von 0,94 g/l wirkte sich fatal auf die Stabilität des Reaktors aus. Nach der Schock-Belastung wurde in der ersten Woche ein kontinuierlicher Abfall im COD- und Phenolabbau beobachtet, gefolgt von einem scharfen Rückgang und der Produktion von Fettsäuren. Eine Erholung des Reaktors war nur möglich, wenn für einen Monat keine Nahrung zugegeben wurde und die Phenolkonzentration beim erneuten Start bei 0,19 g/l gehalten wurde. Die Phenolkonzentration wurde dann allmählich bis auf 1,32 g/l gesteigert, bei verschiedenen OLRs. Bei stabilen Betriebsbedingungen eines Reaktors ohne Glukose im Nährmittel, der vorher mit Glukose und Phenol gefüttert wurde, wurden Zwischenprodukte des anaeroben Phenol-Metabolismus, für einen Zeitraum von 50 Tagen nachdem der totale gelöste COD und der Phenolabbau im Abfluss im Gleichgewicht waren, gefunden.

Im Rahmen dieser Untersuchung wurden sehr hohe 2-CP Abbauraten in anaeroben Anreicherungen von Klärschlamm erhalten. Der Abbau von 50 mg/l 2-CP war in ca. drei Wochen möglich. Die Abbaueffizienz stieg dabei beträchtlich bei fortlaufenden Fütterungen. Der akklimatisierte Schlamm könnte dabei als Inoculum für einen anaeroben Festbettreaktor benutzt werden, um eine hohe 2-CP Abbaurate unter kontinuierlichen Fermentationsbedingungen zu erzielen. Ein stufenweiser Anstieg von 2-CP im Medium erlaubte letztendlich einen erfolgreichen kontinuierlichen Abbau von 2-CP bis zu 2,6 g/l. Das war die höchste getestete zufließende 2-CP Konzentration. Die höchste permanent erreichbare, „sichere" Abbaurate von 2-CP in diesem Reaktor lag um 0,73 g/l·d. Ein negativer Effekt einer 2-CP Schock-Belastung mit 1,36 g/l·d war nach zwei Tagen Hungern und einem Zurücksetzen der 2-CP Belastung reversibel. Innerhalb weiterer zwei Wochen wurde die 2-CP Abbaueffizienz zurückgewonnen und hatte wieder 80 % erreicht. Das weist darauf hin, dass hemmende Effekte aufgrund der hohen Chlorphenolbelastung rückgängig gemacht werden können, wenn diese Belastung kurzzeitig war und den vorherigen Belastungsbedingungen entsprochen wird. Phenol, ein Zwischenprodukt der 2-CP Biodegradation wurde während der Akklimatisierung von Klärschlamm unter Batch-Bedingungen nach jeder Fütterung beobachtet. Während des kontinuierlichen Reaktorbetriebs wurde kein Phenol im Abfluss detektiert. Das weist darauf hin, dass es zu einer schnellen Nutzung des aromatischen Zwischenproduktes kommt.

Organischer Stickstoff und Vitaminquellen wie Pepton und Hefeextrakt können durch NH_4NO_3 und eine definierte Vitaminlösung ersetzt werden, ohne einen dauerhaften negativen Effekt auf die mikrobielle Flora oder den Reaktorbetrieb zu haben. Die durchschnittliche Menge an Chlorid, die pro Einheit 2-CP entfernt wurde lag bei 0,20 g Cl⁻ pro g zugegebenem 2-CP. Das weist auf die erfolgreiche Dechlorierung von 2-CP hin. In einer Batch-Probe mit dem Abfluss vom Reaktor als Inoculum war eine 2-CP Abbaurate von 175 mg/l·d erreichbar. In den Batch-Tests hatte die Anwesenheit von Hefeextract oder Pepton keinen Einfluss auf den Abbau, aber wenn Glukose als Co-Substrat verwendet wurde, sanken die Abbauraten deutlich.

Aus dieser Studie geht hervor, dass die Akklimatisationsstrategie eine wichtige Rolle bei der erfolgreichen biologischen Behandlung von Abwässeren mit toxischen Komponenten wie Phenol und 2-CP in hohen Konzentrationen spielt. Wenn erst einmal eine passende mikrobielle Population etabliert ist, für aeroben/anaeroben Phenol oder 2-CP Abbau, im Biofilm oder in Festbettreaktoren, ist es möglich für Festbettreaktoren schädliche Bedingungen wie Schlamm-Auswaschung, Überladen, Temperaturzusammenbruch, Zusetzen des Reaktors, u.s.w. ohne einen Ausfall des gesamten Prozesses zu überwinden.

References

Abuhamed T., Bayrakter E., Mehmetoglu T., Mehmetoglu U., 2004. Kinetics model for growth of *Pseudomonas putida* F1 during benzene, toluene and phenol biodegradation. Process Biochem.39: 983–988.

Adamo, P. D. D., Rozich, A. F., Daudy, A. F., 1984. Analysis of growth data with inhibitory carbon sources, Biotechnol. Bioeng. 26: 397-402.

Adav S. S., Chen M. Y., Lee D. J., Ren N. Q., 2007. Degradation of phenol by aerobic granules and enrichment yeast *Candida tropicalis*. Biotechnol. Bioeng., 96 (5): 844-852.

Aivasidis A., Diamantis V. I, 2005. Biochemical reaction engineering and process development in anaerobic wastewater treatment. In: T. Scheper (Ed.), Advances in Biochemical Engineering /Biotechnology, Springer, Heidelberg, , Vol 2 : 49-76.

Alexander M., 1998. Biodegradation and bioremediation, second ed., Academic Press, San Diego.

Alexander M., 1994. Biodegradation and bioremediation, first ed., Academic Press, San Diego.

Alexievaa Z., Gerginova M., Zlateva P., Peneva N., 2004. Comparison of growth kinetics and phenol metabolizing enzymes of *Trichosporon cutaneum R57* and mutants with modified degradation abilities. Enzyme Microb. Tech. 34 (3-4): 242-247.

Ambujom S., Manilal V. B., 1995. Phenol degradation by a stable consortium and its bacterial isolates. Biotechnol. Lett. 17 (4): 443-448.

Amor L., Eiroa M., Kennes C., Veiga M. C., 2005. Phenol biodegradation and its effect on the nitrification process. Water Res. 39(13): 2915-2920.

Annachhatre A. P., Gheewala S. H., 1996. Biodegradation of chlorinated phenolic compounds. Biotechnol. Adv. 14, 35-56.

APHA, 1992. Standard Methods for the Examination of Water and Wastewater, 18th ed. American Public Health Association, Washington D. C.

Armenante P. M.,. Fava F., Kafkewitz D., 1995. Effect of yeast extract on growth kinetics during aerobic biodegradation of chlorobenzoic acids. Biotechnol. Bioeng. 47: 227-233.

Armenante P. M., Kafkewitz D., Lewandowski G. A., Jou C. J., 1999. Anaerobic-aerobic treatment of halogenated phenolic compounds. Water Res. 33: 681-692.

Arutchelvan V., Kanakasabai, V., Elangovan, R., Nagarajan, S., Muralikrishnan, V., 2006. Kinetics of high strength phenol degradation using *Bacillus brevis*. J. Hazard. Mater. 129: 216-222.

Arutchelvan, V., Kanakasabai V., Nagarajan, S., Muralikrishnan, V., 2005. Isolation and identification of novel high strength phenol degrading bacterial strains from phenol-formaldehyde resin manufacturing industrial wastewater. J. Hazard. Mater. 127: 238-243.

ATSDR. 1999. Toxicological profile for chlorophenols. U.S. Department of Health and Human Services, Public Health Service. Agency for Toxic Substances and Disease Registry, Atlanta.

ATSDR 1998. Toxicological profile for phenol. U.S. Department of Health and Human Services, Public Health Service. Agency for toxic substances and disease registry, Atlanta.

ATSDR. 2006. Toxicological profile for phenol. U.S. Department of Health and Human Services, Public Health Service. Agency for toxic substances and disease registry, Atlanta.

Bae H. S., Yamagishi T., Suwa, Y., 2002. Evidence for degradation of 2-chlorophenol by enrichment cultures under denitrifying conditions. Microbiol.-SGM 148: 221-227.

Bajaj M., Gallert C. and Winter J., 2008[1]. Anaerobic biodegradation of high strength 2-chlorophenol-containing synthetic wastewater in a fixed bed reactor. Chemosphere. 73 (5) : 705-710.

Bajaj M., Gallert C. and Winter J., 2008[2]. Biodegradation of high phenol containing synthetic wastewater by an aerobic fixed bed reactor. Bioresource Technol. 99 (17): 8376-8381.

Bajaj M., Gallert C. and Winter J., 2008[3]. Effect of co-substrates on aerobic phenol biodegradation by acclimatized and non-acclimatized enrichment cultures. Eng. Life Sci. 8 (2): 125-131.

Bajaj M., Gallert C. and Winter J., 2008[4]. Treatment of phenolic wastewater in an anaerobic fixed bed reactor (AFBR) - Recovery after shock loading. J. Hazad. Mater. doi:10.1016/j.jhazmat.2008.06.027.

Bajaj M., Gallert C. and Winter J., 2008[5]. Biokinetics of phenol degrading culture under aerobic batch conditions. (Communicated).

Bajaj M., Winter J and Gallert C., 2006. Response of shock loading on phenolic wastewater biodegradation in an anoxic suspended bed reactor. In proceedings: 2nd International conference on environmental research and assessment, Bucharest 5-8th, Oct.

Bak F., Widdel, F., 1986. Anaerobic degradation of phenol and phenol derivatives by *Desulfobacterium phenolicum sp. nov.* Arch. Microbiol. 146: 177-180.

Bakker, G., 1977. Anaerobic degradation of aromatic compounds in the presence of nitrate. FEMS Lett. 1: 103-108.

Balch W. E., Fox G. E., Magrum L. J., Woese C. R., Wolfe R. S., 1979. Micorbiol. Rev. 43 (2): 260-296.

Bayly R. C., Barbour M. G., 1984. The degradation of aromatic compounds by the meta and gentisate pathways. In: Microbial Degradation of Organic Compounds. Gibson, D.T. (Ed.), pp. 253-294. Marcel Dekker Inc., New York.

Becker J. G., Stahl D. A., Rittmann B. E., 1999. Reductive dehalogination and conversion of 2-

Chlorophenol to 3-Chlororbenzoate in a methanogenic sediment community: Implications for predicting the environmental fate of chlorinated pollutants. Appl. Environ. Microb. 65 (11): 5169-5172.

Bernet N., Delgenes N., Moletta, R., 1996. Denitrification by anaerobic sludge in piggery wastewater. Environmental Technol. 17: 293–300.

Bertin L., Colao M.C., Ruzzi M., Marchetti L., Fava F., 2006. Performances and microbial features of an aerobic packed-bed biofilm reactor developed to post-treat an olive mill effluent from an anaerobic GAC reactor. Microb. Cell Fact. 5:16.

Beyenal H., Seker S., Tanyolac A., Salih B., 1997. Diffusion coefficients of phenol and oxygen in a biofilm of *Pseudomonas putida*. AICHE J. 43: 243-250.

Blaszczyk M., Przytocka-Jusiak M., Suszek A., Mielcarek, A., 1998. Microbial degradation of phenol in denitrifying conditions. Acta Microbiol. Pol. 47 (1): 65–75.

Bolanos R. M. L., Varesche M. B. A., Zaiat M., Foresti E., 2001. Phenol degradation in horizontal-flow anaerobic immobilized biomass (HAIB) reactor under mesophilic conditions, Water Sci. Technol. 44(4): 167-174.

Bollag J. M., Helling C. S., Alexander M. J., 1986. 2,4-D metabolism. Enymatic hydroxylation of chlorinated phenols. J.Agr. Food Chem. 16: 826-828.

Bradley P. M., 2003. History and ecology of chloroethene biodegradation: a review. Bioremediation J. 7: 81–109.

Buitron G., Gonzalez A., Lopez-Marin, L. M., 1998. Biodegradation of phenolic compunds by an acclimated activated sludge and isolated bacteria. Water Sci. Technol. 37 (4-5): 371-378.

Cha G. C., Chung H. K., Chung J. C., 1997. Suppression of acidogenic activities due to rapid temperature drop in anaerobic digestion. Biotechnol. Lett. 19(5): 461–464.

Chakraborty S., Veeramani H., 2005. Response of pulse phenol injection on an anaerobic-anoxic-aerobic system. Bioresource Technol. 96: 761-767.

Chang C. C., Tseng S. K., Chang C. C., Ho C. M., 2003. Reductive dechlorination of 2-chlorophenol in a hydrogenotrophic gas-permeable, silicone membrane bioreactor. Bioresource Technol. 90: 323-328.

Chang C. C., Tseng S. K., Chang C. C., Ho C. M., 2004. Degradation of 2-Chlorophenol via a hydrogenotrophic biofilm under different reductive conditions. Chemosphere 56: 989-997.

ChemExpo. 1999. Internet site http://chemexpo.com/news/PROFILE990329.cfm.

Chitra S., Sekaran G., Padmavati S., Chandrakasan G., 1995. Removal of phenolic compounds from waste water using mutant strain of *Pseudomonas pictorum*. J. Gen. Appl. Microbiol. 41: 229-237.

Chmielowski J., Grossman A., Labazek S., 1965. Biochemical degradation of some phenols during methane formation. Zesz. Nauk. Polytech. Slask. Inz. Sanit. 8: 97-122.

Cho Y. G., Rhee S. K., Lee S. T., 2000. Influence of phenol on biodegradation of p-nitrophenol by freely suspended and immobilized *Nocardioides sp*. NSP41. Biodegradation. 11: 21-28.

CMAI, 2001. World Cumene/Phenol/Cyclohexane analysis. Chemical Market Associates, Inc. website: www.cmaiglobal.com

CMR 2005. Phenol: Chemical Profile. Chemical Market Reporter, 34-35. May 23-29, 2005.

Cobos-Vasconcelos D .D. L., Santoyo-Tepole F., Juarez-Ramirez C., Ruiz-Ordaz, N., Galindez-Mayer, C. J. J., 2006. Cometabolic degradation of chlorophenols by a strain of *Burkholderia* in fed-batch culture. Enzyme Microb. Tech. 40: 57-60.

Collins G., Foy C., McHugh S., Mahony T., O'Flaherty V., 2005. Anaerobic biological treatment of phenolic wastewater at 15-18^0C, Water Res. 39(8): 1614-1620.

Dietrich G., Winter J., 1990. Anaerobic degradation of chlorophenol by an enrichment culture. Appl. Microbiol. Biot. 34, 253-258.

Dikshitulu S., Baltzis B. C., Lewandowski G. A., Pavlou S., 1993. Competition between two microbial populations in a sequenching fedbatch reactor: Theory, experimental verification, and implications for waste treatment applications, Biotechnol. Bioeng. 42: 643–659.

Dionisi D., Bertin L., Bornoroni L., Capodicasa S., Papini M. P., Fava F., 2006. Removal of organic xenobiotics in activated sludges under aerobic conditions and anaerobic digestion of the adsorbed species. J. Chem. Tech. Biot. 81: 1496-1505.

Donaldson T. L., Strandberg G. W., Hewitt J., Shields G. S., Worden, R. M., 1987. Biooxidation of coal gasification wastewaters using fluidized-bed bioreactors. Environ. Prog. 6: 205-211.

Drioli E., Romano M., 2001. Progress and new perspectives on integrated membrane operations for sustainable industrial growth. Ind. Eng. Chem. Res. 40: 1277-1300.

Eiroa M., Vilar A., Amor L., Kennes C., Veiga M. C., 2005. Biodegradation and effect of formaldehyde and phenol on the denitrification process. Water Res. 39: 449-455.

Erhan E., Yer E., Akay G., Keskinler B., Keskinler D., 2004. Phenol degradation in a fixed bed bioreactor using micro-cellular polymer-immobilized *Pseudomonas syringe*. J. Chem. Technol. Biot. 79: 195-206.

Esparza M. H., Serrano M. C. D., Salinas G. A., Trevino F. A. R., 2006. Removal of high concentrations with adapted activated sludge in suspended form and entrapped in calcium aliginate/cross-linked poly (N-vinyl pyrrolidone) hydrogels. Biotechnol. Progr. 22: 1552-1559.

European Commission, Council Directive 67/548/EEC: Appendix V: Methods for the determination of the Physicochemical Characteristics, the toxicity and the ecotoxicity, 1967. Also available at: http://www.umweltrecht.de/.

Evans W. C. 1947. Oxidation of phenol and benzoic acid by some soil bacteria. Biochem. J. 41: 373-382.

Fakhruddin A. N. M., Quilty B., 2005. The influence of glucose and fructose on the degradation of 2-chlorophenol by *Pseudomonas putida* CP1. World J. Microb. Biot. 21: 1541-1548.

Fang H. H. P., Liang D. W., Zhang T., Liu Y., 2006. Anaerobic degradation of phenol in wastewater under thermophilic condition, Water Res. 40: 427-434.

Fang H. H. P, Liu Y., Ke S. Z., Zhang T., 2004. Anaerobic degradation of phenol in wastewater at ambient temperature, Water Sci. Technol. 49(1): 95-102.

Fang H. H. P, Zhou G. M., 1999. Interactions of methanogens and denitrifiers in degradation of phenols. J. Environ Eng.-ASCE. 125: 57–63.

Farrell A., Quilty B., 1999. Degradation of mono-chlorophenols by a mixed microbial community via a *meta*-cleavage pathway. Biodegradation. 10: 353-362.

Farrell A., Quilty B., 2002. The enhancement of 2-chlorophenol degradation by a mixed microbial community when augmented with *Pseudomonas putida CP1*. Water Res. 36: 2443-2450.

Fedorak P. M., Hrudey S. E., 1988. Anaerobic degradation of phenolic compounds with application to treatment of industrial waste waters. In: Biotreatment Systems, Vol. 1., pp. 170-212. (D. L. Wise, Ed.). CRC Press:Boca Raton, FL.

Feist C. F., Hegeman G. D., 1969. Phenol and benzoate metabolism by *Pseudomonas putida*: regulation of tangential pathways. J. Bacteriol. 100: 869-877.

Field J., Alvarez R. S., 2007. Biodegradability of chlorinated aromatics. In: Science Dossier, Euro-Chlor. Belgium. (Website: www.eurochlor.org).

Flores E. R., Gonzalez M. I., Field J. A., Lora P. O., Grajales L. P., 2003. Biodegradation of mixtures of phenolic compounds in an upward- flow anaerobic sludge blanket reactor. J. Environ. Eng.-ASCE. 129(11): 999-1006.

Gali V. S., Kumar P., Mehrotra I., 2006. Biodegradation of phenol with wastewater as a cosubstrate in upflow anaerobic sludge blanket. J. Environ. Eng.-ASCE 132 (11): 1539-1542.

Galil N., Rebhun M., Brayer Y., 1988. Disturbances and inhibition in biological treatment of wastewater from an integrated refinery. Water Sci. Technol. 3: 21-29.

Gallego A., Gomez C. E., Fortunato M. S., Cenano L., Rossi S., Paglilla M., Hermida H. L.E., Korol S.E., 2001. Factors affecting biodegradation of 2-Chlorophenol by *Alcaligenes sp.* in aerobic reactors. Environ. Toxicol. 4: 306-313.

Gallert C., Winter J., 1992. Comparison of 4-hydroxybenzoate and phenol carboxylase activities in the cell-free extracts of a defined 4-hydroxybenzoate and phenol-degrading anaerobic consortium, Appl. Microbiol. Biot. 37: 119-124.

Gallert C., Winter J., 1993. Uptake of phenol by the phenol-metabolizing bacteria of a stable, strictly anaerobic consortium. Appl. Microbiol. Biot. 39: 627-631.

Gallert C., Winter J., 1994. Anaerobic degradation of 4-hydroxybenzoate: reductive dehydroxylation of 4-hydroxybenzoate by a phenol-metabolizing bacteria of a stable, strictly anaerobic consortium. Appl. Microbiol. Biot. 42: 408-414.

Gallert C., Winter J., 1999. Bacterial metabolism in wastewater treatment systems. In: Rehm H. J., Reed G., Pühler A., Stadler P., Winter J. (Eds.), Biotechnology-Environmental Processes I. Vol. 11 a, second ed., pp. 17-53.Wiley-VCH, Weinheim.

Giorno L., Drioli E., 2000. Biocatalytic membrane reactors: applications and prospectives. Tibtech. 18: 339-349.

Givskov M, Eberi L, Møller S, Poulsen L K, Molin S., 1994 a. Responses to nutrient starvation in *Pseudomonas putida* KT2442: analysis of general cross protection, cell shape, and macromolecular content. J Bacteriol. 176: 7–14.

Givskov M, Eberl L, Molin S., 1994 b Responses to nutrient starvation in *Pseudomonas putida* KT2442: two-dimensional electrophoretic analysis of starvation- and stress-induced proteins. J Bacteriol. 176: 4816–4824.

Goeddertz J. G., Weber A.S., Ying W.C., 1990. Start-up and operation of an anaerobic biological activated carbon (AnBAC) process for treatment of a high strength multicomponent inhibitory wastewater. Environ. Progr. 9: 110-117.

Gonzalez G., Herrera M. G., Garcia M. T., Pena M. M., 2001. Biodegradation of phenol in a continuous process: comparative study of stirred tank and fluidised-bed bioreactors. Bioresource Technol. 76: 245-251.

Goudar C. T., Ganji S. H., Pujar B. G., Strevett K. A., 2000. Substrate inhibition kinetics of phenol biodegradation. Water Environ. Res. 72 (1): 50-55.

Graham J., 2002. Guidance manual for licensing discharges to water. In: European community law. Doc Ref: DLM/COPA/LEG2. pp: 24-37.

Greenan C. M., Moorman T. B., Kaspar T. C., Parkin T. B., Jaynes D. B., 2006. Comparing carbon substrates for denitrification of subsurface drainage. Water Environ Qual. 35: 824-829.

Haggblom M. M., Young, L. Y., 1995. Anaerobic degradation of halogenated phenols by sulfate-reducing consortia. Appl. Environ. Microbiol. 61: 1546-1550.

Hao O. J. Kim M. H., Seagren E. A., Kim H., 2002. Kinetics of phenol and chlorophenol utilization by *Acinetobacter species*, Chemosphere 46 (6): 797-807.

Harwood C.S., Parales R.E., 1996. The β-Ketoadipate pathway and the biology of self-identity. Annu. Rev. Microbiol. 50: 553-590.

He Z., Wiegel J., 1995. Purification and characterization of an oxygen-sensitive 4-hydroxybenzoate decarboxylase from *Clostridium hydroxybenzoicum*. Eur. J. Biochem. 229: 77–82.

Heider J., Fuchs G., 1997. Microbial anaerobic aromatic metabolism, Anaerobe.3(1): 1-22.

Heipieper H.-J., Diefenbach R., Keweloh H., 1992. Conversion of *cis* unsaturated fatty acids to *trans*, a possible mechanism for the protection of phenol-degrading *Pseudomonas putida* P8 from substrate toxicity. Appl. Environ. Microbiol. 58: 1847-1852.

Heipieper H.-J., Weber F. I., Sikkema J., Keweloh H., DeBont I. A. M., 1994. Mechanisms of resistance of whole cells to toxic organic solvents. Trends Biotechnol. 12: 409-415.

Heipieper H.-J., Keweloh H. Rehm J. H., 1990. Influence of phenols on the membrane permeability of free and immobilized bacteria. In: Proceedings of 8th DECHEMA Annual Meeting of Biotechnologists. May 28-29. Behrens, D.; Krämer, P. eds., pp. 641-644, DECHEMA Deutsche Gesellschaft für Chemisches Apparatewesen, New York.

Heipieper H.-J., Keweloh H., Rehm J. H., 1991. Influence of phenols on growth and membrane permeability of free and immobilized *Escherichia coli*. Appl. Environ. Microbiol. 57: 1213-1217.

Hendriksen H. V., Larsen S., Ahring B. K., 1992. Influence of a supplemental carbon source on anaerobic dechlorination of pentachlorophenol in granular sludge. Appl. Environ. Microbiol. 58: 365-370.

Holladay D. W., Hancher C. W., Scott C. D., Chilcote D. D., 1978. Biodegradation of phenolic waste liquors in strirred-tank, packed-bed and fluidized–bed bioreactors, J. Water Pollut. Con. F. 50: 2573-2589.

Hollender J., Hopp J., Dott W., 2000. Cooxidation of chloro- and methylphenols by *Alcaligenes xylosoxidans* JH1. World J. Microb. Biot. 16: 445-450.

Holliger C., Wohlfarth G., Diekert G., 1999. Reductive dechlorination in the energy metabolism of anaerobic bacteria. FEMS Microbiol. Rev. 22: 383-398.

Hsien T. Y., Lin Y.H., 2005. Biodegradation of phenolic wastewater in a fixed biofilm reactor. Biochem. Eng. J. 27: 95-103.

Isken S., DeBont J. A. M. 1996. Active efflux of toluene in a solvent resistance bacterium, J. Bacteriol. 178: 6056-6058.

Jiang H.L., Tay J.H., Tay S.T.L., 2002. Aggregation of immobilized activated sludge cells into aerobically grown microbial granules for the aerobic biodegradation of phenol, Lett. Appl. Microbiol. 35: 439–445.

Jiang, H. L., Tay, J. H., Maszenan, A. M., Tay, S. T. L., 2006. Enhanced phenol biodegradation and aerobic granulation by two coaggregating bacterial strains. Environ. Sci. Technol. 40 (19): 6137-6142.

Kafkewitz D., Fava F., Armenante P. M., 1996. Effect of vitamins on the aerobic degradation of 2-chlorophenol, 4-chlorophenol, and 4-chlorobiphenyl. Appl. Microbiol. Biot. 46: 414-421.

Kar S., Swaminathan T., Baradarajan A., 1996. Studies on biodegradation of a mixture of toxic and non-toxic pollutant using *Arthrobacter species*. Bioprocess Eng., 15: 195-199.

Karigar C., Mahesh A., Nagennahalli M., Yun D. J., 2006. Phenol degradation by immobilized cells of *Arthrobacter citreus*. Biodegradation. 17 (1): 47-55.

Karlsson A., Ejlertsson J., Svensson, B. H., 2000. CO_2-dependent fermentation of phenol to acetate, butyrate and benzoate by an anaerobic, pasteurised culture. Arch. Microbiol. 173: 398-402.

Kazumi J., Haggblom M. M., Young L. Y., 1995. Degradation of monochlorinated and nonchlorinated aromatic- compounds under iron-reducing conditions. Appl. Environ. Microbiol. 61: 4069-4073.

Keith L. H., Telliard W. A., 1979. Priority pollutants I - a perspective view. Environ. Sci. Technol. 13: 416-423.

Keweloh H., Diefenbach R., Rehm, H.-I., 1991. Increase of phenol tolerance of *Escherichia coli* by alterations of the fatty acid composition of membrane lipids. Arch. Microbiol. 157: 49-53.

Keweloh H., Weyrauch G., Rehm J. H., 1990. Phenol-induced membrane changes in free and immobilized *Escherichia coli*. Appl. Microbiol. Biot. 33: 66-71.

Khoury N., Wolfgang D., Kämpfer P., 1992. Anaerobic degradation of phenol in batch and continuous cultures by a denitrifying bacterial consortium. Appl. Microbiol. Biot. 37: 524-528.

Kibret M., Somitsch W., Robra K. H., 2000. Characterization of a phenol degrading mixed population by enzyme assay. Water Res. 34(4): 1127–1134.

Kieboom J., Dennis J. J., DeBont J. A. M., Zylstra G. J., 1998. Identification and molecular characterization of an efflux pump involved in *Pseudomonas putida* S12 solvent tolerance, J. Biol. Chem. 273: 85-91.

Kilby B. A., 1948. The bacterial oxidation of phenol to β-Ketoadipic acid. Proc. Biochem. Soc., Biochem. J. 43: V,VI.

Kim M. H., Hao O. J., 1999. Co-metabolic degradation of chlorophenols by *Acinetobacter species*. Water Res. 33: 562-574.

Knoll G., Winter J., 1987. Anaerobic degradation of phenol in sewage sludge - Benzoate formation from phenol and CO_2 in the presence of hydrogen. Appl. Microbiol. Biot. 25: 384-391.

Knoll G., Winter J., 1989. Degradation of phenol via carboxylation to benzoate by a defined, obligate syntrophic consortium of anaerobic bacteria. Appl. Microbiol. Biot. 30: 318-324.

Kobayashi T., Hashinaga T., Mikami E., Suzuki T., 1989. Methanogenic degradation of phenol and benzoate in acclimated sludges. Water Sci. Technol. 21: 55-65.

Kohring G. W., Rogers J. E., Wiegal J., 1989. Anaerobic biodegradation of 2,4-dichlorophenol in freshwater lake sediments at different temperatures. Appl. Environ. Microb. 55: 348-353.

Krumme M. L., Boyd S. A., 1988. Reductive dechlorination of chlorinated phenols in anaerobic upflow bioreactors. Water Res. 22, 171-177.

Kumar A., Kumar S., Kumar S., 2005. Biodegradation kinetics of phenol and catechol using *Pseudomonas putida* MTCC 1194. Biochem. Eng. J. 22: 151-159.

Kumaran P., Parachuri Y. L., 1997. Kinetics of phenol biotransformation. Water Res. 31: 11-32.

Lack A., Fuchs G., 1994. Evidence that phenol phosphorylation to phenylphosphate is the first step in anaerobic phenol metabolism in a denitrifying *Pseudomonas* sp. Arch. Microbiol. 161: 132–139.

Lallai A., Mura G., Milliddi R., Mastinu C., 1988. Effect of pH on growth of mixed cultures in batch reactor, Biotechnol. Bioeng. 31: 130-134.

Lanthier M., Juteau P., Lepine F., Beaudet R., Villemur R., 2005. *Desulfitobacterium hafniense* is present in a high proportion within the biofilms of a high-performance pentachlorophenol-degrading, methanogenic fixed-film reactor. Appl. Environ. Microb. 71: 1058-1065.

Latkar M., Chakrabarti T., 1994. Resorcinol, catechol and hydroquinone biodegradation in mono and binary substrate matrices in up-flow anaerobic fixed-film fixed–bed reactors. Water Res. 26: 599-607.

Levén L., Nyberg K., Korkea-aho L., Schnürer A., 2006. Phenols in anaerobic digestion processes and inhibition of ammonia oxidising bacteria (AOB) in soil. Sci. Total Environ. 364, 229-238.

Levén L. and Schnürer A. 2005. Effect of temperature on biological degradation of phenols, benzoates and phthalates under methanogenic conditions. Int. Biodeter. Biodegra. 55: 153-160.

Lipthay J.R.D., Barkey T., Vekova J., Sorenson S. J., 1999. Utilization of phenoxyacetic acid, by strains using either the ortho or meta cleavage of catechol during phenol degradation, after conjugal transfer of *tfdA*, the gene encoding a 2,4-dichlorophenoxyacetic acid/2-oxoglutarate dioxygenase. Appl. Microbiol. Biot. 51: 207-214.

Liu Y., Woon K.-H., Yang S.-F., Tay J.-H., 2002. Influence of phenol on cultures of acetate-fed aerobic granular sludge. Letts. Appl. Microbiol. 35: 162–165.

Livingston A. G., Chase H. A., 1989. Modelling phenol degradation in a fluidized bed bioreactor. AICHE J. 35: 1980-1992.

Lob K. C., Tar P. P., 2000. Effect of additional carbon sources on biodegradation of phenol. B. Environ. Contam. Tox. 64: 756-763.

Löffler F. E, Cole J. R, Ritalahti K. M, Tiedje J. M., 2003. Diversity of dechlorinating bacteria. In: Microbial Processes and Environmental Applications. Haggblom MM, Bossert ID, eds. pp. 53–87. Kluwer. Boston/Dordrecht/London.

Loh K. C., Wang S. J., 1998. Enhancement of biodegradation of phenol and a non-growth substrate 4-chlorophenol by medium augmentation with conventional carbon sources. Biodegradation. 8: 329-338.

Loh K. C., Wu T. T., 2006. Cometabolic transformation of 2-chlorophenol and 4-chlorophenol in the presence of phenol by *Pseudomonas putida*. Can. J. Chem. Eng. 84: 356-367.

Lovley D. R., Lonergan D. J., 1990. Anaerobic oxidation of toluene, phenol and p-cresol by the dissimilatory iron-reducing organism, GS-15. Appl. Environ. Microb. 56: 1858-1864.

Majumder P. S., Gupta S. K., 2007. Removal of chlorophenols in sequential anaerobic-aerobic reactors. Bioresource Technol. 98: 118-119.

Margesin R., Bergauer P., Gander S., 2004. Degradation of phenol and toxicity of phenolic compounds: a comparison of cold-tolerant *Arthrobacter sp.* and mesophilic *Pseudomonas putida*. Extremophiles. 8(3): 201-207.

Marrot B., Martinez A. B., Moulin P., Roche N., 2006. Biodegradation of high phenol concentration by activated sludge in an immersed membrane bioreactor. Biochem. Eng. J. 30: 174-183.

Millar G. L., 1959. Use of dinitrosalicylic acid reagent for determination of reducing sugar. Anal. Chem. 31: 426–428.

Moersen A., Rehm H. J., 1990. Degradation of phenol by a defined mixed culture immobilized by adsorption on activated carbon and sintered glass. Appl. Microbiol. Biot. 33: 206-212.

Monteiro A. A. M. G., Boaventura R. A. R., Rodrigues A. E., 2000. Phenol biodegradation by *Pseudomonas putida* DSM 548 in a batch reactor. Biochem. Eng. J. 6: 45-49.

Mulcahy L. T., Shieh W. K., Motta E. J. L., 1980. Kinetic model of biological fluidized bed biofilm reactor (FBBR). Prog. Water Technol. 12: 143-158.

Nikaido H., 1994. Prevention of drug access of bacterial targets: permeability barriers and active efflux, Science. 264: 382-388.

Nordin K., Unell M., Jansson J. K., 2005. Novel 4-chlorophenol degradation gene cluster and degradation route via hydroxyquinol in *Arthrobacter chlorophenolicus* A6. Appl. Environ. Microb. 71: 6538-6544.

Nuhoglu A., Yalcin B., 2005. Modelling of phenol removal in a batch reactor. Process Biochem. 40 (3–4): 1233–1239.

Pallerla S., Chambers R. P., 1998. Reactor development for biodegradation of pentachlorophenol. Catal. Today. 40: 103-111.

Papanastasiou A. C., Maier W. J., 1992. Kinetics of biodegradation of 2,4-dichlorophenoxyacetate in the presence of glucose. Biotechnol. Bioeng. 14: 2001-2011.

Pawlosky U., Howell J. A., 1973. Mixed culture biooxidation of phenol- Determination of kinetic parameters. Biotechnol. Bioeng. 15: 889–896.

Pinto, G., Pollio, A., Previtera, L.,Stanzionel, M. Temessi, F., 2003. Removal of low molecular weight phenols from olive oil mill wastewater using microalgae. Biotechnol. Lett. 25: 1657–1659.

Przybolewska K., Wieczorek A., Nowak A., Pochrzaszcz M., 2006. The isolation of microorganisms capable of phenol degradation. Pol. J. Microbiol. 55 (1): 63-67.

Quan, X. C., Shi H. C., Zhang Y. M., Wang J. L., Qian Y., 2003. Biodegradation of 2,4-dichlorophenol in an air-lift honeycomb-like ceramic reactor. Process Biochem. 38: 1545-1551.

Richards D. J, Shieh W. K., 1986. biological fate of organic priority pollutants in aquatic environment. Water Res. 20(9): 1077-1090.

Roslev P., King G. M., 1995. Aerobic and anaerobic starvation metabolism in methanotrophic bacteria. Appl. Environ. Microb. 61 (4): 1563-1570.

Rozich A. F., Colvin R. J., 1985. Effects of glucose on phenol biodegradation by heterogeneous populations. Biotechnol. Bioeng. 28: 65-71.

Rutgers M., Breure A. M., van Andel J. G., Duetz W.A., 1997. Growth yield coefficients of *Sphingomonas* sp. strain P5 on various chlorophenols in chemostat culture. Appl. Microbiol. Biot. 48: 656-661.

Saravanan P., Pakshirajan K., Saha P., 2008. Growth kinetics of an indigenous mixed microbial consortium during phenol degradation in a batch reactor. Bioresource Technol. 99: 205-209.

Sarafaraz S., Thomas S., Tewari U. K., Iyengar L., 2004. Anoxic treatment of phenolic wastewater in the sequencing batch reactor. Water Res. 38: 965-971.

Satsangee R., Ghosh P., 1990. Anaerobic degradation of phenol using an acclimated mixed culture. Appl. Microbiol. Biot. 34: 127-130.

Schie, P. M. V, Young L. Y., 2000. Biodegradation of phenol: Mechanisms and applications. Bioremediation J. 4(1):1–18.

Sharma A., Kachroo D., Kumar R., 2002. Time dependent influx and effect of phenol by immobilized microbial consortium. Environ. Monit. Assess. 76: 195-211.

Shen D. S., He R., Liu X. W., Long Y., 2006. Effect of pentachlorophenol and chemical oxygen demand mass concentrations in influent on operational behaviours of upflow anaerobic sludge blanket (UASB) reactor. J. Hazard. Mater. 136: 645-653.

Shetty K.V., Ramanjaneyulu R., Srinikethan G., 2007. Biological phenol removal using immobilized cells in a pulsed plate bioreactor: effect of dilution rate and influent phenol concentration. J. Hazard. Mater. 149 (2): 452-459

Sittig M. 1975. Environmental Sources and Emissions Handbook, Noyes Data Corporation, Park Ridge, NJ.

Smith M. H., Woods S.L., 1994. Regiospecificity of chlorophenol reductive dechlorination by vitamin B_{12}. Appl. Environ. Microb. 60: 4111-4115.

Solyanikova I. P., Golovleva L. A., 2004. Bacterial degradation of chlorophenols: Pathways, biochemica and genetic aspects. J. Environ. Sci. Heal. B. 39: 333-351.

Song B. K., Palleroni N. J., Häggblom M. M., 2000. Isolation and characterization of diverse halobenzoate-degrading denitrifying bacteria from soils and sediments. Appl. Environ. Microb. 66: 3446–53.

Sontheimer H., Spindler P., Rohmann U.,1986. Wasserchemie für Ingenieure. Scriptum Engler-Bunte-Institut Universität Karlsruhe (TH) 185 S.

Spain J. C., Gibson D. T, 1988. Oxidation of substituted phenols by *Pseudomonas putida* F1 and *Pseudomonas* sp. Strain JS6. Appl. Environ. Microb. 54 (6): 1399–1404.

Stainer R. Y., Pallerom N. J., Doudorott M., 1966. The aerobic *Pseudomonas*: A taxonomic study. J. Gen. Microbiol. 43: 159-271.

Stanier R. Y., Ornston L. N., 1973. The β-Ketoadipate pathway. Adv. Microb. Physiol. 9: 89-151.

Tartakovsky B., Manuel M. F., Beaumier D., Greer C. W., Guiot S. R., 2001. Enhanced selection of an anaerobic pentachlorophenol-degrading consortium. Biotechnol. Bioeng. 73: 476-483.

Tawfiki K., Lepine F., Bisaillon J., Beaudet R., 1999. Simultaneous removal of phenol, *o-* and *p*-cresol by mixed anaerobic consortia, Can. J. Microbiol. 45: 318-325.

Tawfiki K., Lepine F., Bisaillon J.-G., Beaudet R., Hawari J., Guiot S.R., 2000. Effects of bioaugumentation strategies in UASB reactors with methanogenic consortium for removal of phenolic compounds, Biotechnol. Bioeng. 67 (4): 419-423.

Tay S.T.L., Moy B.Y.P., Jiang H.L., Tay J.H., 2005. Rapid cultivation of stable aerobic phenol-degrading granules using acetate-fed granules as microbial seed, J. Biotechnol. 115: 387-395.

Tay J. H., Jiang H. L., Tay S. T. L., 2004. High-rate biodegradation of phenol by aerobically grown microbial granules. J. Environ. Eng.-ASCE. 130(12): 1415–1423.

Tay J. H., Liu Q. S., Liu Y., 2001 a. Microscopic observation of aerobic granulation in sequential aerobic sludge blanket reactor. J. Appl. Microbiol. 91(1): 168–175.

Tay J. H; He Y. X., Yan Y. G., 2001 b. Improved anaerobic degradation of phenol with supplemental glucose. J. Environ. Eng.-ASCE. 127: 38-45.

Tiedje J. M., 1988. Ecology of denitrification and dissimilatory nitrate reduction to ammonium.. In A.J.B. Zehnder (ed.) Biology of anaerobic microorganisms. pp. 179–244. Wiley-Interscience, Ontario.

Tiedje J. M., Sexstone A. J., Myrold D. D., Robinson J.A., 1982. Denitrification: Ecological niches, competition and survival. Antonie van Leeuwenhoek 48: 569–583.

Tziotzios G., Teliou M., Kaltsouni V., Lyberatos G., Vayenas D. V., 2005. Biological phenol removal using suspended growth and packed bed reactors. Biochem. Eng. J. 26: 65-71.

U.S. EPA. 1979. Phenol. Ambient water criteria, PB 296 787. office of water planning and standards, United states environmental protection agency, Washington, DC.

Uygur A., Kargir F., 2004. Phenol inhibition of biological nutrient removal in a four-step sequencing batch reactor. Process Biochem. 39: 2123-2128.

Van Overbeek L. S., Eberi L., Givskov M., Molin S., Van Elsas J. D., 1995. Survival of, and induced stress resistance in, carbon-starved *Pseudomonas fluorescens* cells residing in soil. Appl Environ Microb. 61: 4202–4208.

Van Schie, P. M., Younq, L. Y., 2000. Biodegradation of Phenol: Mechanisms and applications. Bioremediation. J. 4(1): 1-18.

Vazquez I., Rodriguez J., Maranon E., Castrillon L., Fernandez Y., 2006. Simultaneous removal of phenol, ammonium and thiocyanate from coke wastewater by aerobic biodegradation. J. Hazard. Mater. 137(3): 1773-1780.

Villemur R., Lanthier M., Beaudet R., Lepine F., 2006. The *Desulfitobacterium* genus. FEMS Microbiol. Rev. 30: 706-733.

Wang Y. S., Barlaz M .A., 1998. Anaerobic biodegradability of alkylbenzenes and phenol by landfill derived microorganisms. FEMS Microbiol. Ecol. 25: 405-418.

Wang Y. T., Shanmuganathan M., Wang Z., 1998. Reductive dechlorination of chlorophenols in methanogenic cultures. J. Environ. Eng.-ASCE. 124: 231-238.

Waste water technology: origin, collection, treatment and analysis of waste water, 1989. Deutsche Gesellschaft für Technishe Zusammenarbeit (GTZ) GmbH. Springer-Verlag, Berlin Heidelberg.

Watanabe K., Hino S., Takahashi N., 1996. Responses of activated sludge to an increase in phenol loading. J. Ferment. Bioeng., 82(5): 522–524.

Watanabe K, Miyashita M., Harayama S., 2000. Starvation improves survival of bacteria introduced into activated sludge. Appl. Environ. Microb. 66(9): 3905–3910.

Weber A. S., Lai M.-S., Lin W., Goeddertz J. G., Ying W.-C., 1992. Anaerobic/aerobic biological activated carbon (BAC) treatment of a high strength phenolic wastewater. Environ. Progr. ENVPDI. 11(4): 310-317.

Wei X, Bauer W D., 1998. Starvation-induced changes in motility, chemotaxis, and flagellation of *Rhizobium meliloti*. Appl Environ Microb. 64: 1708–1714.

WHO, 1989. Chlorophenols other than pentachlorophenol. Environmental Health Criteria. World Health Organization, Geneva, Switzerland.

Wolf P., Nordmann W., 1977. A study-method for measuring COD of wastewaters. Korrespondenz Abwasser (In German). 24: 277-281.

Worden R. M., Donaldson T. L., 1987. Dynamics of a biological fixed film for phenol degradation in a fluidized-bed bioreactor. Biotechnol. Bioeng. 30: 398–412.

Wrangstadh M, Szewzyk U, Ostling J, Kjelleberg S., 1990. Starvation-specific formation of a peripheral exopolysaccharide by a marine *Pseudomonas* sp., strain S9. Appl Environ Microb. 56: 2065–2072.

Yan J., Jianping W., Hongmei L., Suliang Y., Zongding H., 2005. The biodegradation of phenol at high initial concentration by the yeast *Candida tropicalis*. Biochem. Eng. J. 24 (3): 243-247.

Yang C. F., Lee C. M., Wang C. C., 2005. Degradation of chlorophenols using pentachlorophenol-degrading bacteria *Sphingomonas chlorophenolica* in a batch reactor. Curr. Microbiol. 51: 156-160.

Ye. F. X., Shen D. S., 2004. Acclimation of anaerobic sludge degrading chlorophenols and the biodegradation kinetics during acclimation period. Chemosphere 54: 1573-1580.

Yu J., Ward O. P., 1994. Studies on factors influencing the biodegradation of pentachlorophenol by a mixed bacterial culture. Int. Biodeter. Biodegr. 33: 209-221.

Yum K. J., Peirce J. J., 1998. Continuous-flow wood chip reactor for biodegradation of 2,4-DCP. J. Environ. Eng.-ASCE 124: 184-190.

Zache G., Rehm H. J., 1989. Degradation of phenol by a co-immobilized entrapped mixed culture, Appl. Microbiol. Biot. 30: 426-432.

Zhang X., Wiegal J., 1990. Sequential anaerobic degradation of 2,4-dichlorophenol in freshwater sediments. Appl. Environ. Microb. 56: 1119-1127.

Zheng Y.-M., Yu H.-Q., Liu S.-J, Liu, X.-Z. 2006. Formation and instability of aerobic granules under high organic loading conditions. Chemosphere. 63 (10): 1791-1800.

Zhou G., Fang H., 1997. Co-degradation of phenol and *m*-cresol in an UASB reactor, Bioresource Technol. 61: 47-52.

Zilouei H, Guieysse B., Mattiasson B., 2006. Biological degradation of chlorophenols in packed-bed bioreactors using mixed bacterial consortia. Process Biochem. 41: 1083-1089.

Appendix I

Picture 1: Aerobic fixed bed reactor for phenolic wastewater

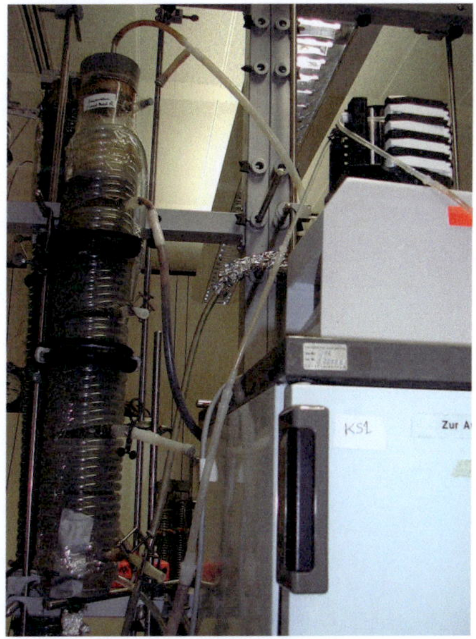

Picture 2: Anaerobic fixed bed reactor for phenolic wastewater

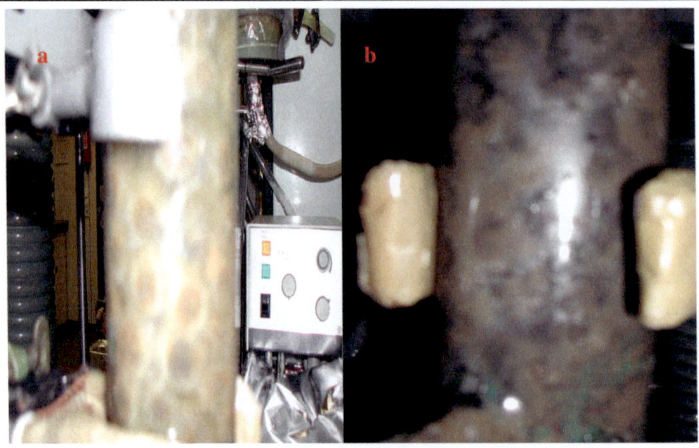

**Picture 3: Sludge growth on aerobic fixed bed 3 weeks (a) and
6 weeks after reactor start-up (b)**

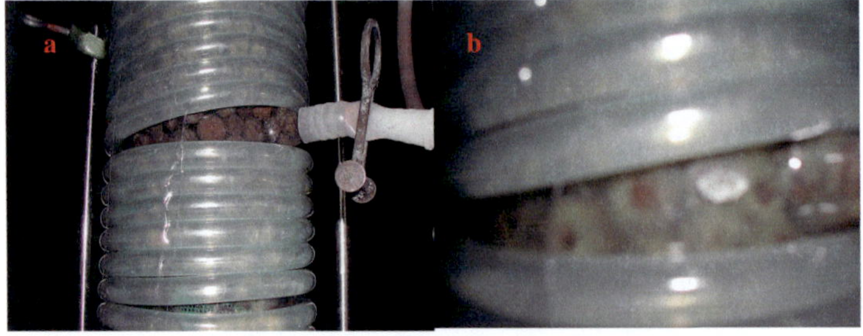

**Picture 4: Sludge growth on anaerobic fixed bed 3 weeks (a) and
6 weeks after reactor start-up (b)**

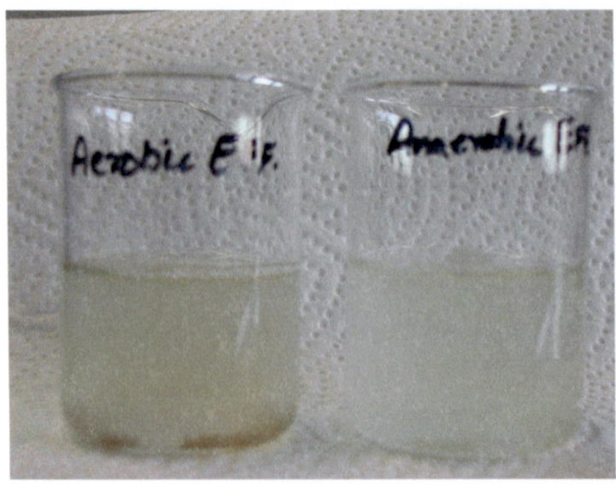

Picture 5: A comparison between aerobic and anaerobic fixed bed effluents

Picture 6: Anaerobic suspended bed reactor for phenolic wastewater

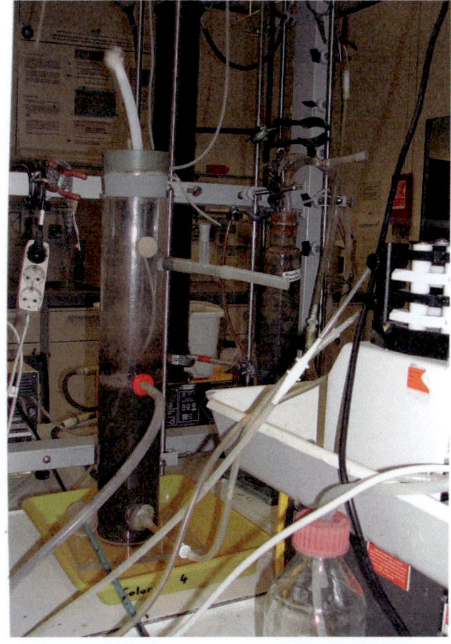

Picture 7: Aerobic suspended bed reactor for phenolic wastewater

Picture 8: Anoxic suspended bed reactor for phenolic wastewater

Picture 9: Anaerobic suspended bed reactor for 2-CP degradation

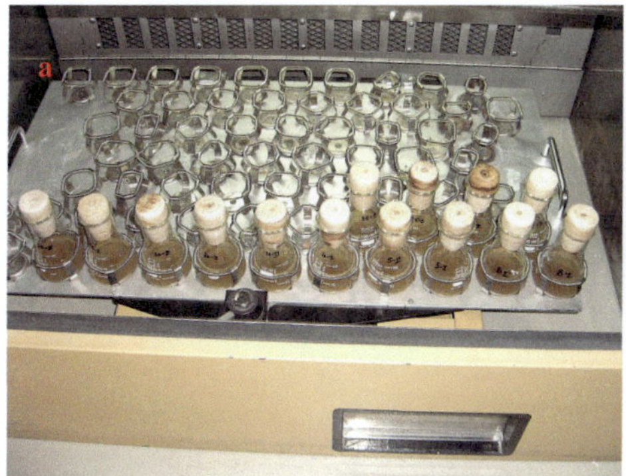
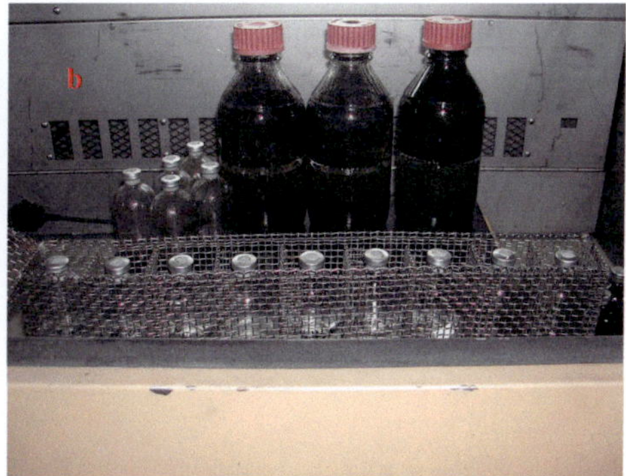

Picture 10: Aerobic (a) and anaerobic (b) batch assays

**Picture 11: Anoxic batch (a) and acclimatisation of anaerobic sludge
to 2-CP under batch conditions (b)**

Picture 12: Aerobic batch assays with enrichment culture obtained after several sludge transfers at the start of experiment (a) and biomass growth after 18 hours at different phenol concentrations (b)

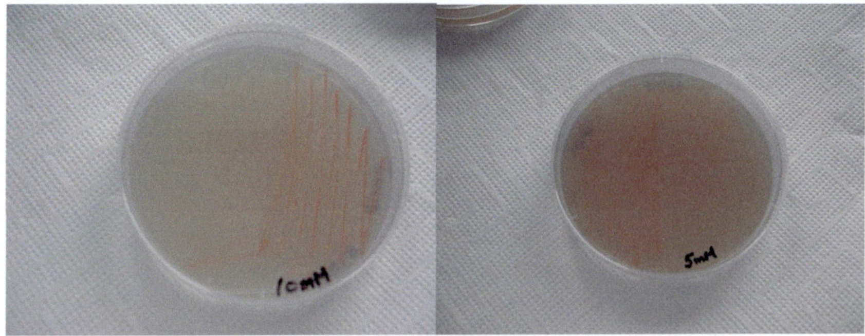

Picture 13: Growth of isolated aerobic culture on agar plates at different phenol concentrations

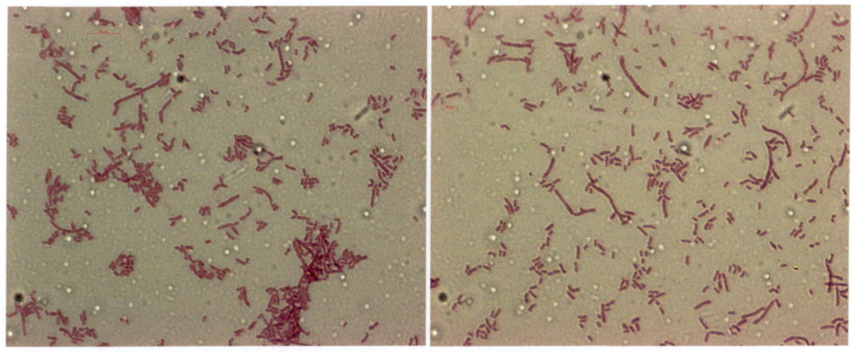

Picture 14: Pure strain of bacteria isolated from aerobic fixed bed reactor and identified as *Rhodococcus sp.* (1000X)

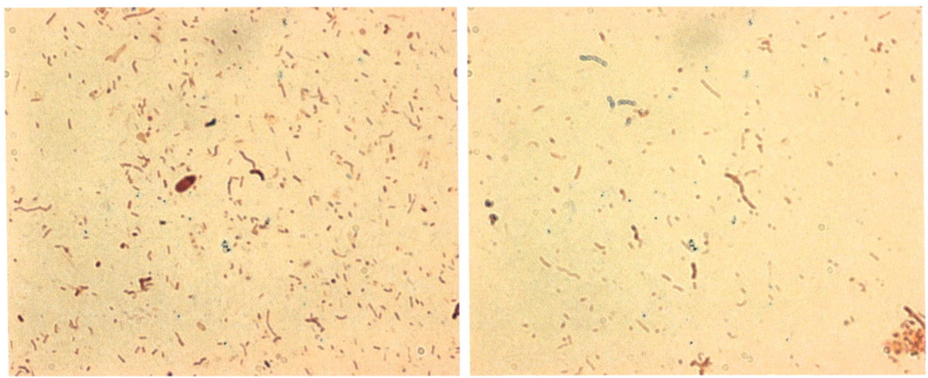

Picture 15: Microbial flora in anoxic reactor (1000 X)

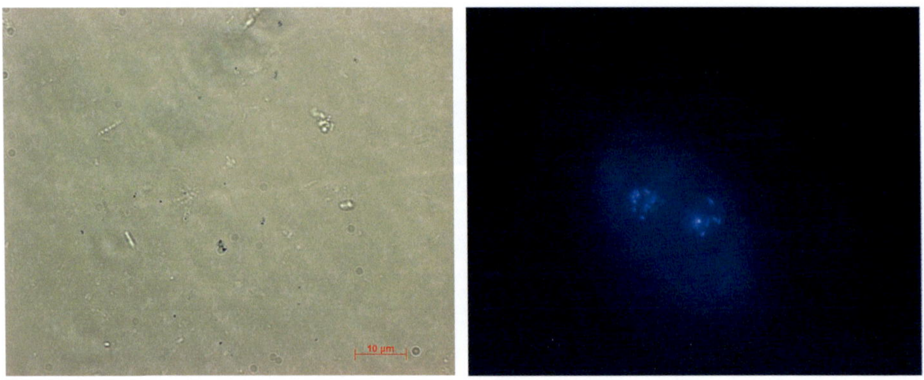

Picture 16: Phase contrast and UV-microscopy of the microbial flora present in the effluent of anaerobic fixed bed reactor (1000 X)

Appendix II

Standard curves

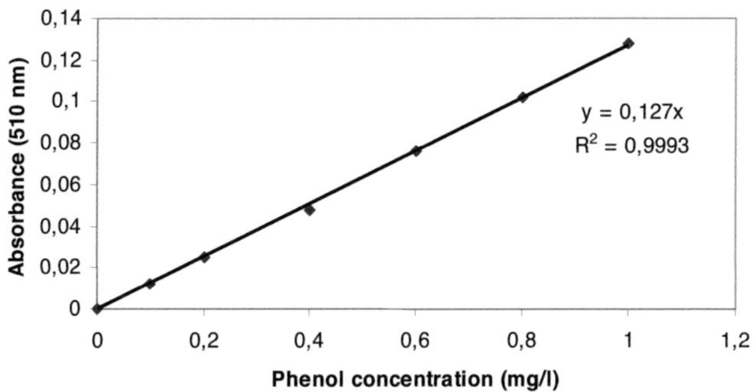

Fig. 1 – Phenol standard calibration curve from concentrations 0-1 mg/l in deionised water

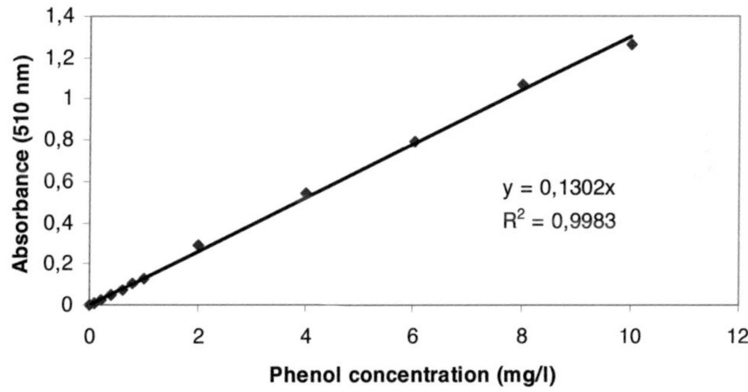

Fig. 2 – Phenol standard calibration curve from concentrations 0-10 mg/l in deionised water

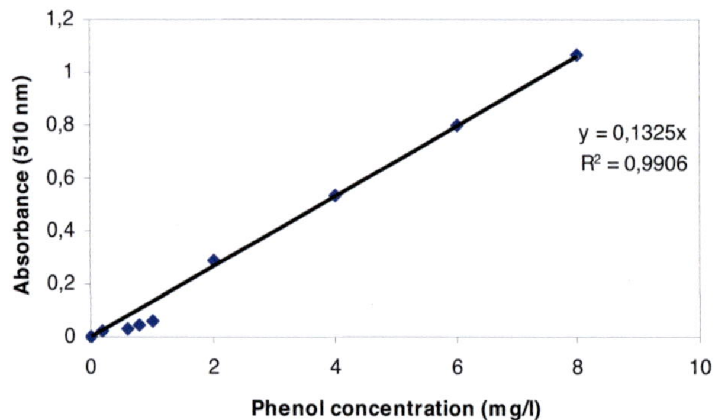

Fig. 3 – Phenol calibration curve from concentrations 0-8 mg/l in reactor influent medium

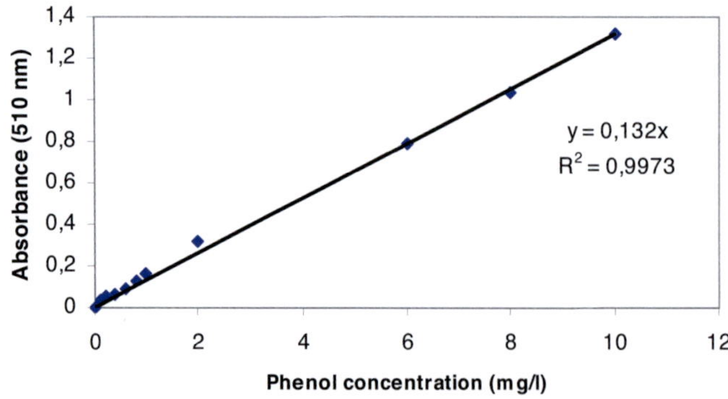

Fig. 4 – Phenol calibration curve from concentrations 0-10 mg/l in reactor effluent

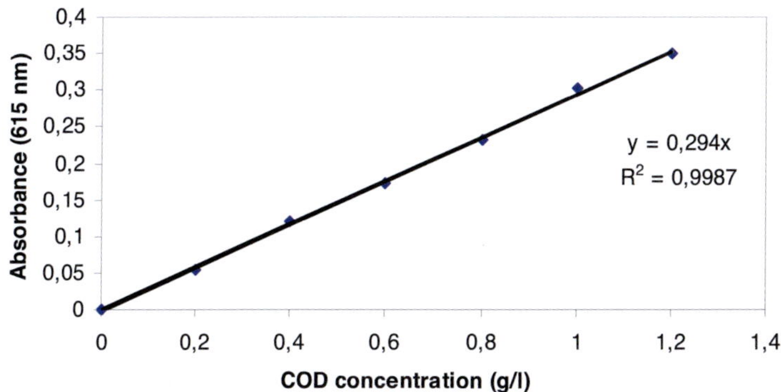

Fig. 5 – COD calibration curve for concentrations 0-1.2 g/l

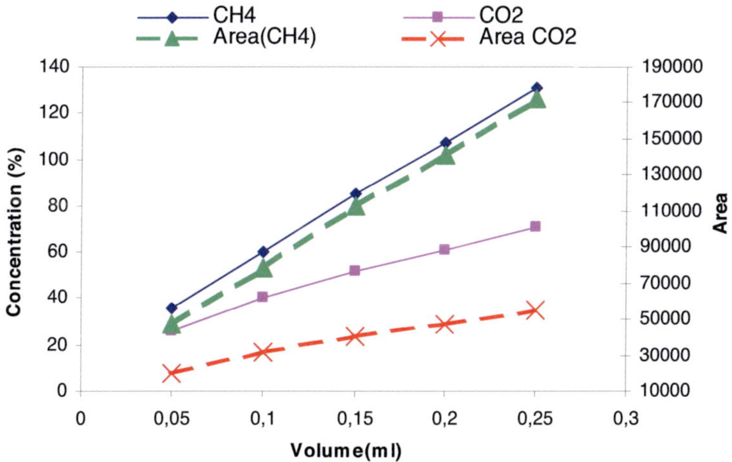

Fig. 6 – Biogas calibration curve

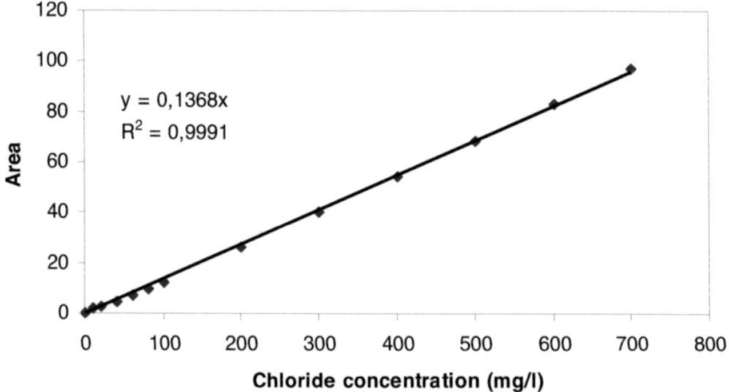

Fig. 7 – Chloride standard curve

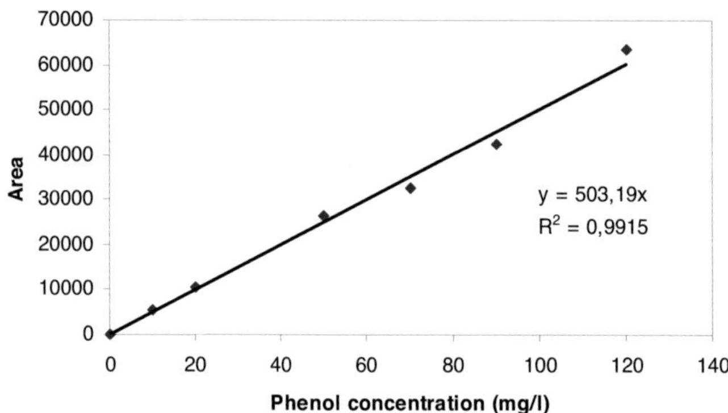

Fig. 8 – Phenol standard curve for gas chromatograph

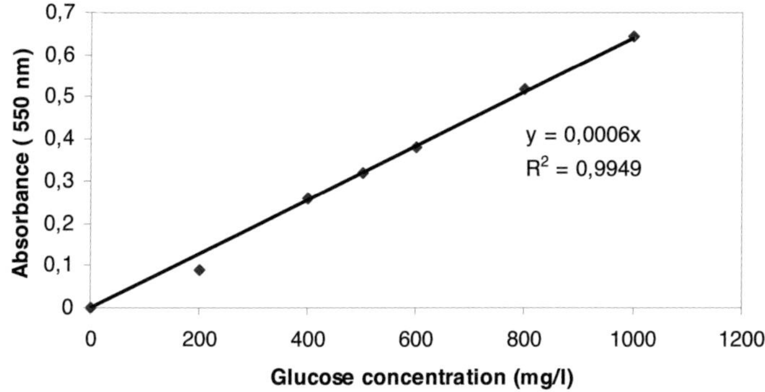

Fig. 9 – Glucose standard curve

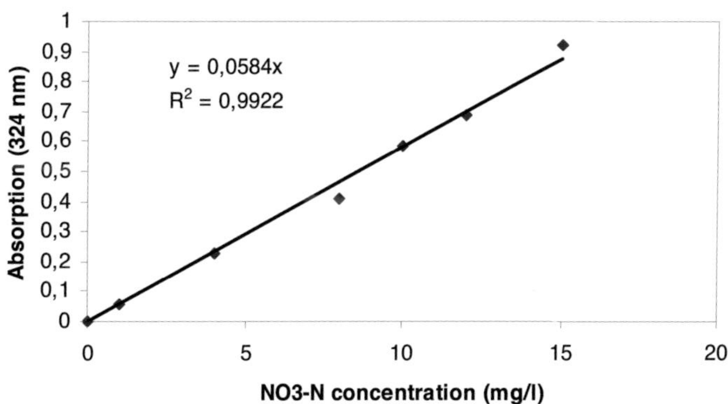

Fig. 10 – Nitrate standard curve

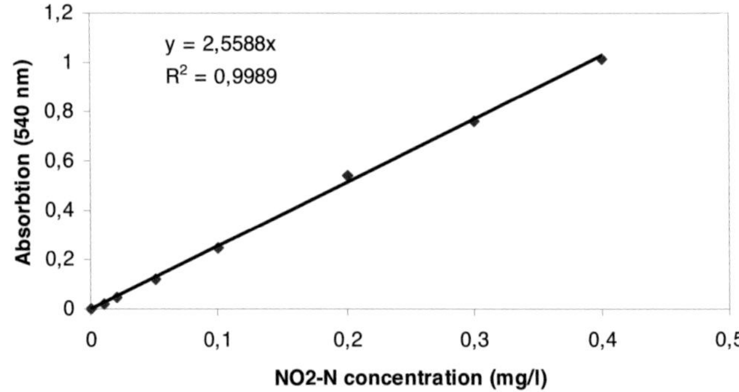

Fig. 11 – Nitrite standard curve

Treatment of phenol and 2-chlorophenol containing wastewaters in bioreactors

Mini Bajaj

Institut für Ingenieurbiologie und
Biotechnologie des Abwassers

Universität Karlsruhe (TH)
Forschungsuniversität · gegründet 1825

Presentation outline

- Background information about phenol and 2-chlorophenol (2-CP)

- Aims of the present research work

- Research approach and outcome

- Conclusions

Phenol and Chlorophenols (CP)

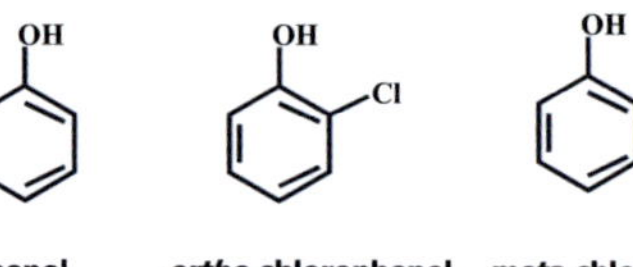

Phenol	*ortho*-chlorophenol	*meta*-chlorophenol	*para*-chlorophenol
	(2-CP)	(3-CP)	(4-CP)

- Aromatic compounds
- Five types of chlorinated phenols: mono-,di-,tri-,tetra- and penta-chlorophenols (19 in total)
- Mono-chlorophenols- 2CP, 3CP and 4-CP

Applications of phenol

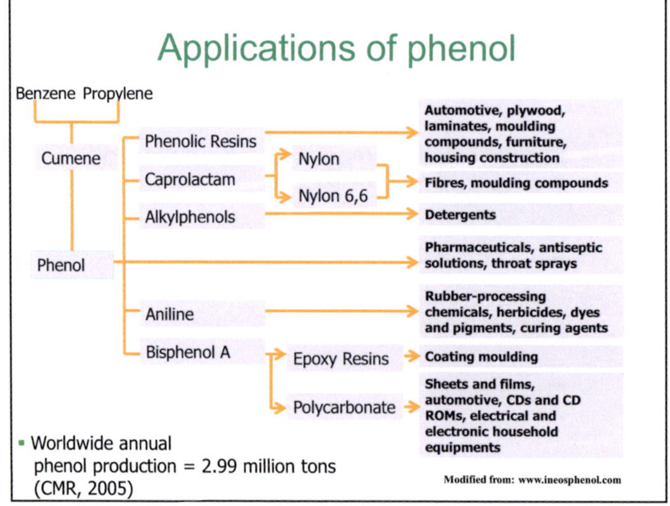

- Worldwide annual phenol production = 2.99 million tons (CMR, 2005)

Levels of phenol in industrial wastewaters

Industrial Source	Phenol concentration (mg/l)
Coke Ovens	
Weak ammonia liquor without dephenolization	600-12,000
Weak ammonia liquor with dephenolization	4-330
Oil Refineries	
Low-temperature carbonization effluent	3395
Sour water	80-185
General waste stream	10-100
Mineral oil wastewater	100
Petrochemicals	
General wastewater	50-600
Tar distilling plants	300
Herbicide manufacturing	210
Other	
Plastics factory	600-2,000
Phenolic resin production	15,000
Dephenolization liquour	3,000
Stocking production	6,000
Fiberglass manufacturing	40-400

Source: Environmental sources and emissions handbook, NJ

Anthropogenic sources of chlorophenols

- Fungicides and herbicides

- Wood preservative

- By-product in pulp and paper bleaching

- Degradation intermediates of some xenobiotics

Phenol and 2-CP exposure and toxicological effects

- Contaminated food and water
- Strong odour and taste (in ppm to ppb)
- EU classification
  - Toxic
  - Harmful
  - Corrosive
  - Dangerous for environment
- USEPA- Priority pollutant
- Microbial inhibition may lead to **partial or entire biological waste water treatment plant failure**

Bactericidal effects of phenol/2-CP

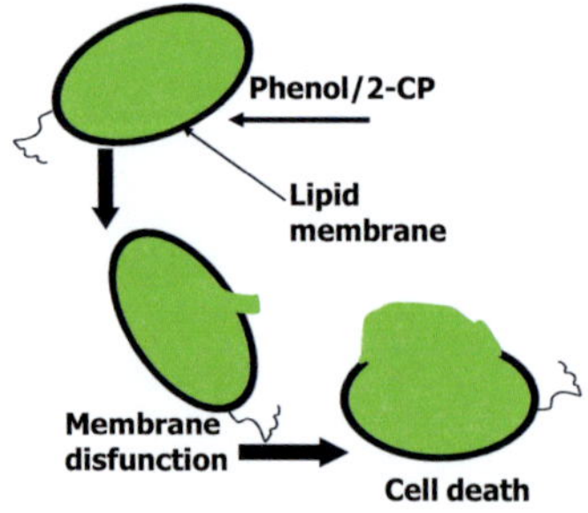

- Phenol/2-CP dissolve in the cytoplasmic membrane and disrupt its function e.g. by
  - Leakage of K^+ ions
  - Leakage of ATP/ nucleotides at bactericidal concentration

Source- Heipieper et al. (1991)

Mechanism of acclimatisation of aerobic bacteria to phenol

- Re-establishment of membrane functions after non-lethal toxicant exposure e.g. by
  - Increase in protein/lipid ratios
  - Adapted flora is able to reabsorb K^+ ions

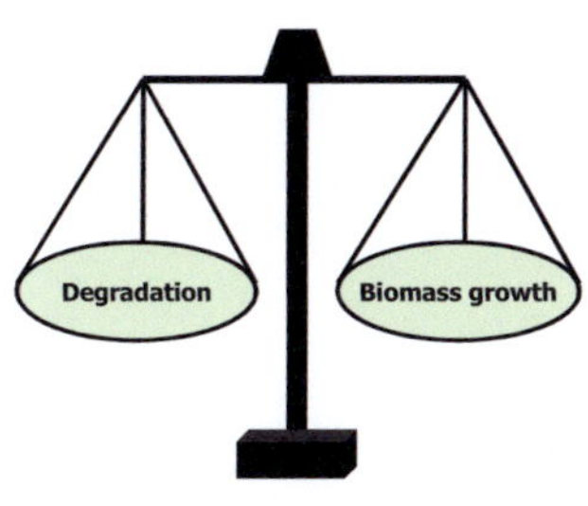

Source : Heipieper et al. (1991)

Patterns of biological phenol removal

Oxygen respiration

$$C_6H_5OH + 7\,O_2 \longrightarrow 6\,CO_2 + 3\,H_2O$$

Nitrate respiration

$$5\,C_6H_5OH + 28\,NO_3^- + 28\,H^+ \longrightarrow 14\,N_2 + 30\,CO_2 + 29\,H_2O$$

Methanogenesis

$$C_6H_5OH + 4\,H_2O \longrightarrow 3.5\,CH_4 + 2.5\,CO_2$$

Literature studies

- Most biodegradation studies deal with 0.47 to 1.4 g phenol/l or 0.01 to 0.26 g 2-CP/l

➔ Problems with handling of higher concentrated phenolic influents

Research objectives

- To demonstrate the ability of a microbial consortium of domestic sewage sludge to acclimatise to toxic compounds

- To biodegrade high concentrations of phenol or 2-CP in wastewater

- To compare aerobic, anoxic and anaerobic treatment of wastewater in suspended biomass and biofilm reactors for phenol or 2-CP removal

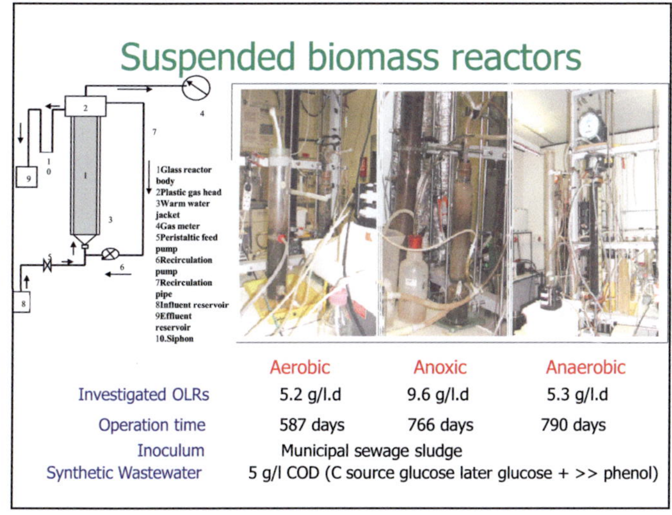

	Aerobic	Anoxic	Anaerobic
Investigated OLRs	5.2 g/l.d	9.6 g/l.d	5.3 g/l.d
Operation time	587 days	766 days	790 days
Inoculum	Municipal sewage sludge		
Synthetic Wastewater	5 g/l COD (C source glucose later glucose + >> phenol)		

Reactor performance in terms of phenol removal

Suspended biomass reactor	Max. phenol removal rate (g/l·d)	Phenol concentration (g/l)	Loading rate (g/l·d)		Operation time with phenol (d)
			Total (COD)	Phenol	
Anoxic	0.75	0.75	5.90	0.84	522
Aerobic	0.55	0.75	4.13	0.65	488
Anaerobic	0.44	1.13	2.18	0.48	647

1 g Phenol = 2.38 g COD

- Degradation under anoxic conditions is better than under aerobic and anaerobic conditions

- Measured removal rates are in the range of previous studies

➔ Therefore further experiments with biofilm reactors

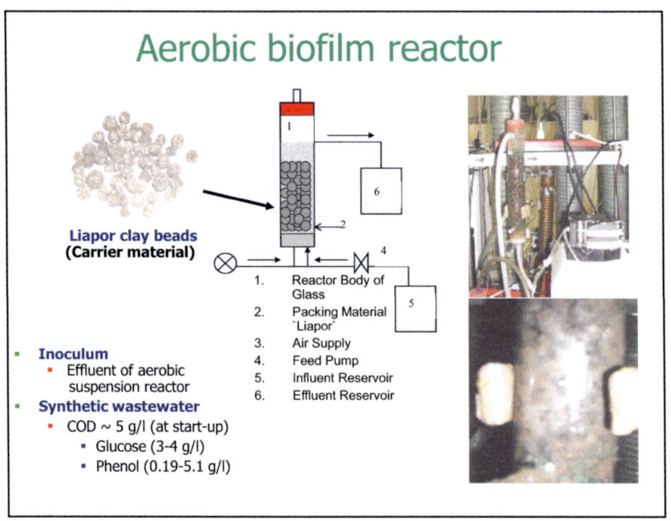

Aerobic biofilm reactor
Liapor clay beads
(Carrier material)
1
6
2
4
5
1. Reactor Body of Glass
2. Packing Material `Liapor`
3. Air Supply
4. Feed Pump
5. Influent Reservoir
6. Effluent Reservoir
Inoculum
Effluent of aerobic suspension reactor
Synthetic wastewater
COD ~ 5 g/l (at start-up)
Glucose (3-4 g/l)
Phenol (0.19-5.1 g/l)

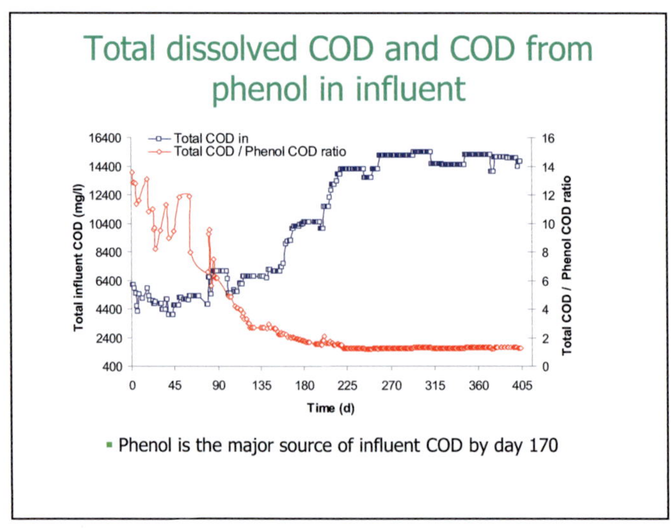

Total dissolved COD and COD from phenol in influent
Total influent COD (mg/l)
Total COD / Phenol COD ratio
Total COD in
Total COD / Phenol COD ratio
16400
14400
12400
10400
8400
6400
4400
2400
400
16
14
12
10
8
6
4
2
0
0
45
90
135
180
225
270
315
360
405
Time (d)
Phenol is the major source of influent COD by day 170

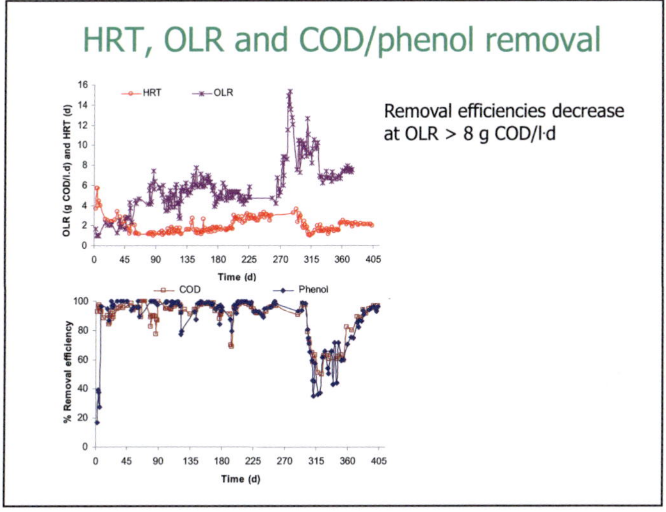

Phases during reactor operation

- Reactor started at 1.67 g COD/l·d OLR
- Within 2 months OLR increased to 4.5 g/l·d
- Maximum OLR tested 15.3 g COD/l·d, 'safe' OLR ~ 8 g COD/l·d
- Steady state removal of influent COD of 14.9 g/l
- Influent phenol concentration
    - At reactor start up: 0.19 g/l
    - Max. conc. tested: 5.2 g/l
    - Steady state phenol removal: 4.9 g/l (= 11.7 g/l COD)

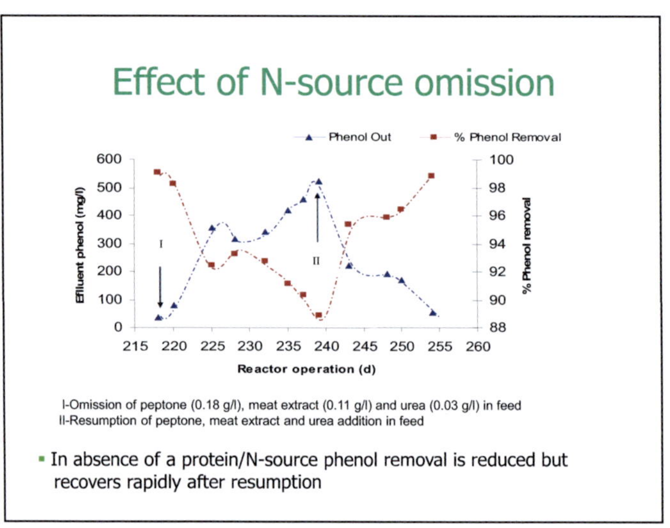
Effect of N-source omission
Phenol Out
% Phenol Removal
Effluent phenol (mg/l)
% Phenol removal
600
500
400
300
200
100
0
100
98
96
94
92
90
88
215 220 225 230 235 240 245 250 255 260
Reactor operation (d)
I
II
I-Omission of peptone (0.18 g/l), meat extract (0.11 g/l) and urea (0.03 g/l) in feed
II-Resumption of peptone, meat extract and urea addition in feed
In absence of a protein/N-source phenol removal is reduced but recovers rapidly after resumption

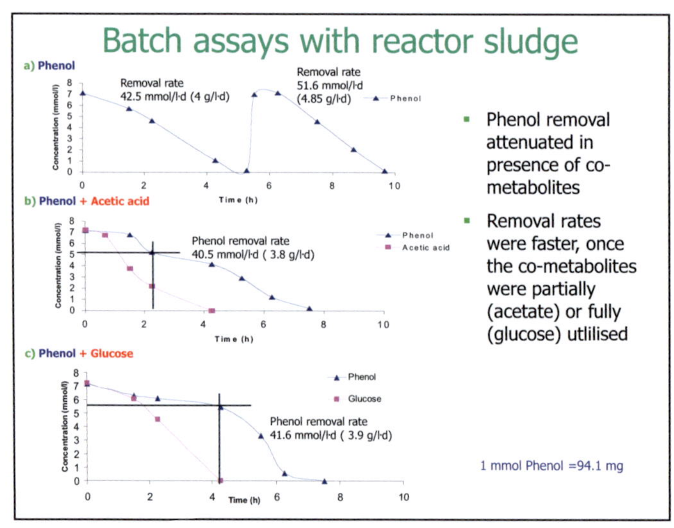
Batch assays with reactor sludge
a) Phenol
Concentration (mmol/l)
8 7 6 5 4 3 2 1 0
Removal rate
42.5 mmol/l·d (4 g/l·d)
Removal rate
51.6 mmol/l·d
(4.85 g/l·d)
Phenol
0 2 4 6 8 10
Time (h)
b) Phenol + Acetic acid
Concentration (mmol/l)
8 7 6 5 4 3 2 1 0
Phenol removal rate
40.5 mmol/l·d (3.8 g/l·d)
Phenol
Acetic acid
0 2 4 6 8 10
Time (h)
c) Phenol + Glucose
Concentration (mmol/l)
8 7 6 5 4 3 2 1 0
Phenol
Glucose
Phenol removal rate
41.6 mmol/l·d (3.9 g/l·d)
0 2 4 Time (h) 6 8 10
Phenol removal attenuated in presence of co-metabolites
Removal rates were faster, once the co-metabolites were partially (acetate) or fully (glucose) utlilised
1 mmol Phenol =94.1 mg

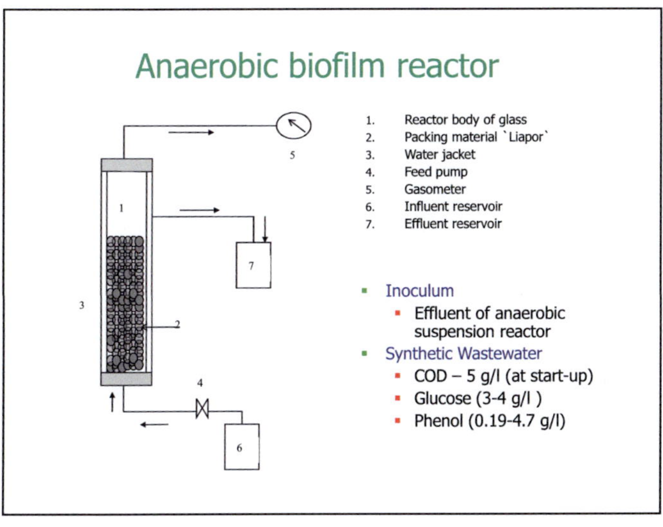
Anaerobic biofilm reactor
1. Reactor body of glass
2. Packing material `Liapor`
3. Water jacket
4. Feed pump
5. Gasometer
6. Influent reservoir
7. Effluent reservoir
Inoculum
Effluent of anaerobic suspension reactor
Synthetic Wastewater
COD – 5 g/l (at start-up)
Glucose (3-4 g/l)
Phenol (0.19-4.7 g/l)

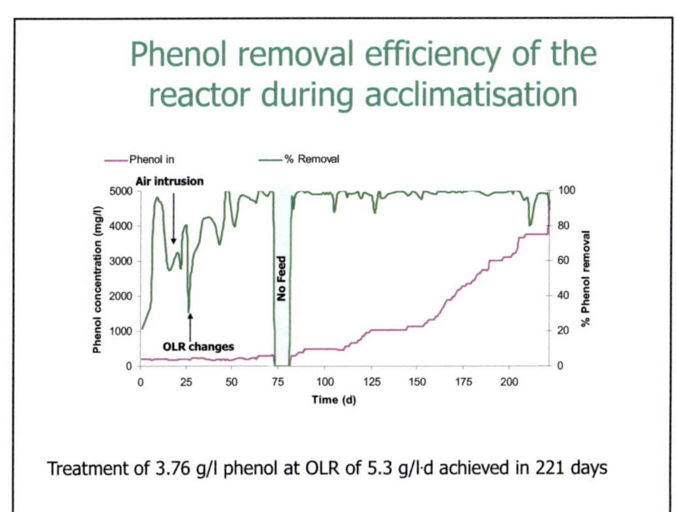
Phenol removal efficiency of the reactor during acclimatisation
Phenol in
% Removal
Air intrusion
Phenol concentration (mg/l)
% Phenol removal
No Feed
OLR changes
5000
4000
3000
2000
1000
0
100
80
60
40
20
0
0 25 50 75 100 125 150 175 200
Time (d)
Treatment of 3.76 g/l phenol at OLR of 5.3 g/l·d achieved in 221 days

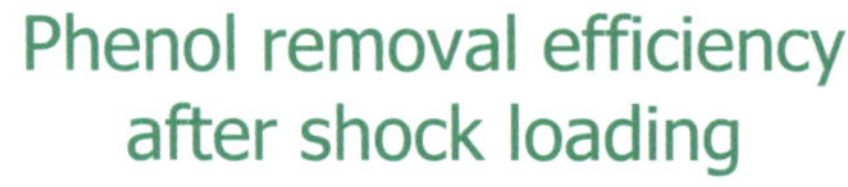

- An increment of the phenol concentration from 3.76 g/l to 4.7 g/l at day 222 had a very negative effect
- After a period of minimal/no feed, system started to recover

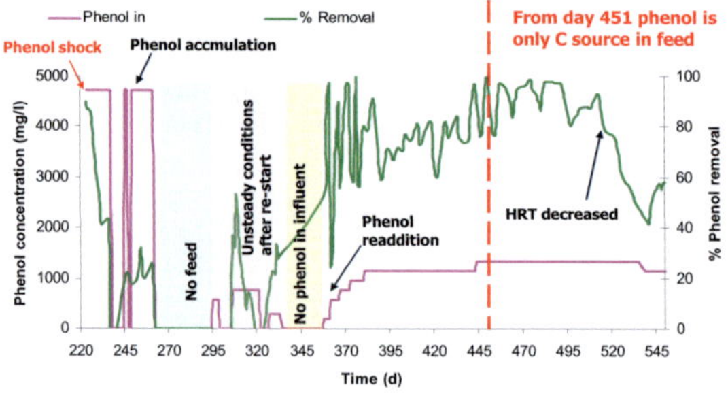

Comparison of phenol shock in anaerobic systems with present study

References / Parameters	Gali et al., 2006	Fang et al., 2004	Bajaj et al., 2008 (present study)
System	UASB	UASB	AFBR
Operation time (d)	632	~336	550
Substrate	Phenol + Molasses	Phenol	Phenol + Glucose (after 451 only phenol)
Phenol conc.(g/l) (before shock loading)	1.17	-	3.76
Max. phenol conc. (g/l)	1.34	1.26	4.76
Max. COD conc. (g/l)	4.0	~3.0	14.2 (Total), 11.2 (from phenol)
Maximum OLR (g COD/l.d)	8.0	18.0	5.7
Removal (%) (before shock load)	~ 88-99.9 (COD)	95 (Phenol)	93.4 (COD), 98.5 (Phenol)
Period of shock load (d)	~20	5	15
Remarks	Complete reactor failure	6.7 months for recovery	Partial recovery in 4 months

Biodegradation of 2-CP

- Acclimatisation of municipal sewage sludge to 2-CP up 50 mg/l in batch reactors
- Start-up of a continuous anaerobic biofilm reactor with acclimatised inoculum
- Continuous increase in 2-CP influent concentration up to 2.6 g/l
- Batch experiments

Sludge acclimatisation to 2-CP in fed-batch reactor (first 3 feeding cycles)

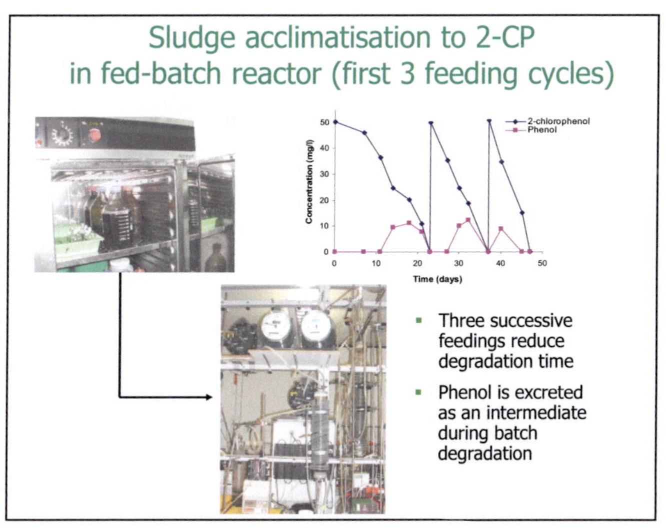

- Three successive feedings reduce degradation time
- Phenol is excreted as an intermediate during batch degradation

Removal efficiency of 2-CP at increased loading rates (LR) in an anaerobic biofilm reactor

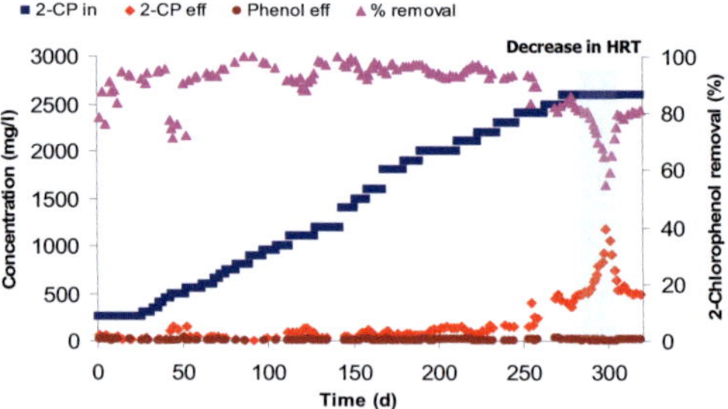

- Sludge pre-acclimatisation enabled reactor start up with a relatively high 2-CP concentration (0.25 g/l)
- The maximum 2-CP LR was 1.36 g 2-CP/l·d at 1.9 d HRT

Decreasing removal efficiencies with increasing 2-CP loading rate

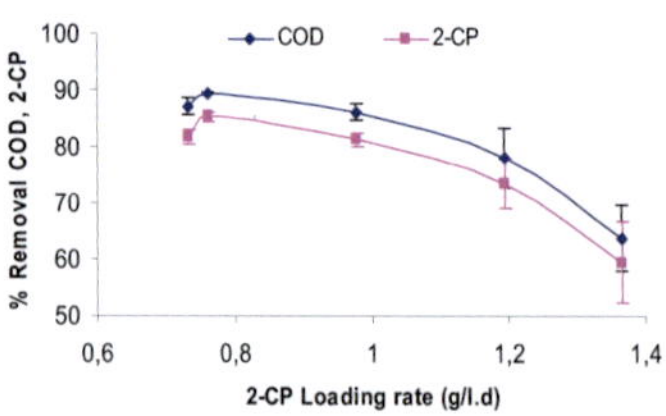

- For 2-CP loading rates up to 1 g/l·d, percent removal of 2-CP is between 80 to 85%
- For a 2-CP loading rate of 1.36 g/l·d removal decreases to < 60%

Performance of various systems for 2-CP degradation

Reactor	2-CP Conc. (mg/l)	2-CP Loading rate 2-CP mg/l·d	2-CP Removal rate (mg/l·d)	Remarks	Reference
Suspended batch	-	-	180	Enrichment culture	Dietrich & Winter (1990)
AFBR[a]	2000	70-600	375	Phenol accmulation observed	Dietrich & Winter (1990)
SMBR[b]	25	40	37.1	Hydrogenotrophic conditions	Chang et al. (2003)
Batch	12.8	-	-	Denitrifying conditions	Bae et. al. (2002)
Batch	10-25	-	about 0.3 (maximum)	Experiment carried out in presence and absence of sucrose	Ye and Shen (2004)
UASB[c]	30	45-120	111.6	3 g/l sodium acetate used as co-substrate	Majumder and Gupta (2007)
Suspended batch	50-192	-	**114.7-175.2**	Experiment carried out with acclimatised inoculum	Bajaj et al. (2008) **Present study**
AFBR	**250-2600**	50-1365	**873**	At the end of reactor operation removal rate was 613 mg/l.d	Bajaj et al., (2008) **Present Study**

a-AFBR- Anaerobic fixed bed reactor, b-SMBR- Silicone membrane bioreactor, c-UASB-Upflow anaerobic sludge bed

Maximum performance parameters obtained in the presented study

Reactor type	Max. removal rate (g/l·d)	Substrate		OLR (COD g/l.d)		HRT (d)
		Type	Conc. (g/l)	Total	From phenol/2CP	
Biofilm: Aerobic	2.93	Phenol	4.90	15.30	12.92	0.95
Biofilm: Anaerobic	1.62	Phenol	4.70	5.68	4.47	2.50
Suspension: Anoxic	0.75	Phenol	0.75	5.90	2.0	0.90
Suspension: Aerobic	0.55	Phenol	0.75	4.13	1.54	1.16
Suspension: Anaerobic	0.44	Phenol	1.13	2.18	1.15	2.34
Biofilm: Anaerobic	0.87	2-CP	2.60	2.33	1.91	2.20

Conclusions

- Phenol degradation in suspended biomass reactors was best under anoxic conditions

- For wastewater containing high concentrations of phenolic compounds, biofilm reactors are more efficient
  - High removal rates of high phenol/2-CP concentrations in wastewater could be achieved by stepwise increase of OLR
  - Aerobic systems show higher resistance and faster recovery in response to shock loadings than anaerobic systems
  - Nitrogen omission from the influent slightly decreased removal efficiency in biofilm reactors but would effect the efficiency of suspended biomass reactors much more

- 2-CP is more toxic than phenol and acclimatisation of sewage sludge to 2-CP degradation was possible with 50 mg/l
  - 2.6 g 2 CP/l was the highest treated influent concentration ever tested
  - Recovery after shock loading at 1.36 g 2-CP/l·d was possible

Acknowledgements

I thank DAAD, Bonn – The German Academic Exchange Service for providing me a PhD grant

Thank you for your kind attention !

Publications

1. **Bajaj M.**, Gallert C. and Winter J. (2008) "Effect of co substrates on aerobic phenol biodegradation by acclimatized and non-acclimatized enrichment cultures." Eng. Life Sci. 8(2): 125-131.

2. **Bajaj M.**, Gallert C. and Winter J. (2008). "Biodegradation of phenol containing synthetic wastewater in an aerobic fixed bed reactor" Bioresource Technol.99(17): 8376-8381. doi:10.1016/j.biortech.2008.02.057

3. **Bajaj M.**, Gallert C. and Winter J. (2008) "Treatment efficiency of anaerobic fixed bed reactor (AFR) at increasing phenol loading and its possible recovery after shock loading." J. Hazard. Mater. doi:10.1016/j.jhazmat.2008.06.027

4. **Bajaj M.**, Gallert C. and Winter J. (2008)"Anaerobic biodegradation of high strength 2-Chlorophenol containing synthetic wastewater in a fixed bed reactor". Chemosphere 73(5): 705-710 doi:10.1016/j.chemosphere.2008.06.072

5. **Bajaj M**, Gallert C and Winter J (2006). "Response of shock loading on phenolic wastewater biodegradation in an anoxic fluidised bed reactor": In proceedings of 2nd International Conference on Environmental Research and Assessment, Bucharest 5-8th Oct.

6. **Bajaj M.**, Gallert C. and Winter J. (2008) "Biokinetics of phenol degrading culture under aerobic batch conditions" **(Communicated)**